ADVANCE COPY

This advance copy is one of a limited number being distributed solely for review and examination purposes to selected persons without charge. Orders cannot be delivered prior to the publication date in January of the copyright year, when this publication is offered for general sale.

HOLT, RINEHART AND WINSTON, INC.

Statistical
Analysis

Books by Allen L. Edwards

Experimental Design in Psychological Research, *Fourth Edition*
Statistical Methods, *Third Edition*
Expected Values of Discrete Random Variables and Elementary Statistics
The Social Desirability Variable in Personality Assessment and Research
  Techniques of Attitude Scale Construction
Probability and Statistics
The Measurement of Personality Traits by Scales and Inventories
Statistical Analysis, *Fourth Edition*

FOURTH EDITION

Statistical Analysis

Allen L. Edwards

Professor of Psychology, University of Washington

HOLT, RINEHART AND WINSTON, INC.
New York   Chicago   San Francisco   Atlanta
Dallas   Montreal   Toronto   London   Sydney

Library of Congress Cataloging in Publication Data

Edwards, Allen Louis.
  Statistical analysis.

  First ed. published in 1946 under title:
Statistical analysis for students in psychology and education.
  1. Statistics. 2. Educational statistics.
I. Title.
HA29.E3   1974   519   73-9503
ISBN 0-03-006901-7

Preface

Despite the recognition accorded statistical methods by active workers in psychology, education, and the behavioral sciences, some undergraduate majors in these fields regard the customary course in statistics as both dull and frustrating. And there is no denying the fact that their actual experience in the course may serve to reinforce rather than to change these beliefs.

Even though a student may look forward to a course in statistics with some degree of interest, this interest is often dulled before the course is completed. Loss of interest, I believe, often results from the stress that is frequently placed on long and involved problems that are essentially exercises in multiplication, division, addition, and subtraction. Most students tend to regard these problems — and perhaps rightfully so — as so much "busy work." And the course in statistics will undoubtedly prove to be frustrating to those students who do not have the degree of mathematical training that the instructor assumes they have.

Under the conditions described, it is not surprising that some students' memories of the statistics course are memories of mysterious derivations (which they failed to understand) and of laborious calculations (which they were compelled to perform). It is these memories that are passed on to the members of next semester's class.

This text attempts to break the vicious circle. The illustrations that I have used in the text are, in general, of such a nature that *any* student can follow the arithmetic easily and without the necessity of wasting time in attempts to trace the course of a series of involved calculations. In addition, I have tried to present elementary statistical theory in a way that will be intuitively meaningful and reasonable to the nonmathematically trained student.

I have followed the same plan in the questions and problems that appear at the end of each chapter. Instead of requiring laborious computations, the exercises are quite simple. They are designed to aid the student, providing a review of the material in the text and enabling the student to test his understanding of the text and also his ability to apply what he has learned.

Some exercises are designed to show the student that a principle or theorem illustrated in the text with one set of data also applies to other sets of data. Still other exercises are designed so that the student can discover for himself a theorem or principle. To the nonmathematically trained student, discovering that a principle works or "discovering" the principle itself is much more satisfying than being presented with a proof that he does not understand. Having discovered that a principle works, he may then be ready for a general proof.

I have included some proofs in the text and in the answers to the problems but, in general, they are not emphasized. The reader or instructor who desires a text in which proofs are emphasized should consider one of my more advanced books, *Probability and Statistics* or *Statistical Methods*, Third Edition. These two books contain the proofs of almost everything that the reader of this text is asked to accept as a result of a demonstration.

Students, of course, should know that demonstrations are not proofs. The motivated and capable student may wish to attempt to develop his own proofs of the theorems and principles presented in the text or to consult either of the two books mentioned above to see the proofs given in them.

In this, the fourth, edition of *Statistical Analysis*, I have deleted some sections that appeared in the third edition, added a number of new sections, and have rewritten many other sections in an attempt to improve clarity. Some changes have also been made with respect to the organization of the sections within chapters. In addition, I have added a considerable number of new problems and exercises for each of the chapters. In all cases where a problem asks for a numerical answer or a proof, these are given in the Answers to the Problems at the back of the book.

I am indebted to the late Professor Sir Ronald A. Fisher, and to his publishers Messrs. Oliver and Boyd Limited, Edinburgh, for permission to reprint Table VI from their book *Statistical Methods for Research Workers*. Table V has been reproduced from Professor George W. Snedecor's book *Statistical Methods* by permission of the author and his publisher, the Iowa State College Press. Table IV has been reprinted from Enrico T. Federighi: Extended Tables of the Percentage Points of Student's t Distribution, *Journal of the American Statistical Association*, 1959, Volume 54, Number 287, pp. 683–688, by permission. Table I is reproduced by permission of M. G. Kendall and B. B. Smith and the Royal

Statistical Society. Portions of Table II have been taken from *Handbook of Statistical Nomographs, Tables, and Formulas* by permission of J. W. Dunlap and A. K. Kurtz.

To various instructors and students who have informed me of their reactions to the third edition of *Statistical Analysis*, I am also indebted. In this respect, I am particularly indebted to Professor Henry F. Kaiser of the University of California, Berkeley.

Allen L. Edwards

Seattle, Washington
October 1973

Contents

Statistical
Analysis

1

Introduction

1.1 The Importance of Statistics for Psychologists and Teachers

In many departments of psychology and in many colleges of education, a course in statistical methods is required of all students as part of their major sequence. If statistics is a required course for you, it is because your advisers believe that an understanding of elementary statistics is essential for every psychologist and teacher, regardless of your other major interests.

If you are a prospective clinical or social psychologist, you may feel that it is all right to require statistics for those students whose interest is in experimental psychology, but not for you. What do statistical methods have to do with clinical or social psychology? That they have a great deal to do with the subject matter of these two fields of psychology can easily be demonstrated. The next time you are in the library, select any issue of the *Journal of Consulting and Clinical Psychology*, the *Journal of Personality and Social Psychology*, the *Journal of Abnormal Psychology*, or the *Journal of Counseling Psychology*, and glance through it. How many of the articles can you read and understand without any knowledge of statistical methods? How many contain references to means, standard deviations, variances, and correlation coefficients? How many mention tests of significance? Do you find symbols such as t, F, and χ^2 used in these articles? If you do not understand elementary statistics, you will not be able to read and understand the research reported in the journals devoted to your specialty.

If you are a prospective teacher of English or social science rather than a prospective psychologist, you may feel that it is all right to require statistics for psychologists, but what does it have to do with the teaching of English or social science? Perhaps not too much with the actual teaching of the subject matter, but a great deal with the profession of teaching. Research takes place in education as well as in psychology. You will need to know something of elementary statistics if you are to understand reports of educational research. To convince yourself that this is so, examine any issue of the *Journal of Educational Psychology*, the *Journal of Educational Measurement*, or the *American Educational Research Journal*. These journals report the results of research studies of interest to teachers. To be able to read and understand the articles published in these journals you will need to know something about statistical methods.

Regardless of whether you are in the process of becoming a psychologist or a teacher, you will need to understand elementary statistics in order to understand and interpret scores on various kinds of psychological and educational tests. Much of the work of the clinical psychologist involves the administration and interpretation of psychological tests. In modern school systems, standardized achievement tests are administered routinely. As a teacher, you should be able to understand and interpret scores on these tests. To do this, you must be able to read and understand test manuals. A test manual describes the manner in which a test is developed and standardized, and this process involves a great deal of statistics.

1.2 The Text and the Student

Approached from the point of view that *statistical techniques are tools* to be used in experimentation and research, and in the discovery of new facts, the study of statistical methods can be a very interesting as well as valuable subject. As psychologists and teachers, are we interested in descriptions? Then statistical methods can assist us in making our descriptions more precise. Are we interested in differences between individuals and groups? Then statistical methods can assist us in describing and evaluating such differences. Are we interested in discovering whether there is any relationship between two traits or between performances on two different tests? Statistical methods again come to our assistance. These are applications of statistical methods to problems of interest to teachers and psychologists. In reading this text, you should constantly keep in mind the potential applications of the techniques described to problems of interest to you. How can you use these techniques? What would they help you to find out about some of those problems?

Statistical techniques may be regarded as techniques for classifying, describing, and evaluating data. To learn to understand the logical and theoretical basis of elementary statistical techniques does not require any elaborate previous mathematical training. The field of mathematical statistics, of course, is highly developed, and not every worker in the field of psychology or education can be expected to be an expert in statistics as well as in his subject-matter specialty. This does not mean, however, that the subject-matter specialist cannot gain an *understanding* of the basis of statistical techniques. Once he has gained this understanding, he can then read journal articles that report the results of research in his field of specialization. In addition, he can apply statistical methods to problems of interest to him, with some feeling of confidence that his use of these methods is not inappropriate.

In this text we shall attempt, by elementary methods, to show the person with little mathematical training something of the basis of statistical techniques. The procedures to be followed in making various computations are shown in detail. Emphasis is placed on the interpretation of the results of these computations, along with any assumptions that may be involved in the process. It is possible to do this just as well with numbers that are small as with numbers that are large. For this reason, simple examples are given so that computations can be readily followed and understood. This is an instructional device, and you should keep in mind that the real world is seldom as simple as the examples described.

A book written about the subject matter of statistical methods and a course in statistical methods may not be quite like the usual texts and courses to which you are accustomed. Some courses do not require daily preparation, and many students get into the habit of waiting until just before an examination before getting down to work. By cramming, they may succeed in absorbing a sufficient amount of knowledge, temporarily at least, to pass an objective or essay examination. It is well known that material learned in this way is soon forgotten. Students may not consider this too great a handicap, if they find that an understanding of the topics covered later in the text or course is not dependent on what has come before.

This is not the case with statistical methods. They cannot be successfully learned or mastered by cramming. Nor can the student, once having taken an examination, afford to forget the material studied and still expect to understand what is to come later. Statistical methods, as presented in this book, start from scratch: the assumption is that the student knows nothing at all about the subject. But there is a continuity of development, with each new topic or section building on the foundation established in the beginning. Constant review of material previously covered, until you have mastered it, is essential.

1.3 Symbols and Formulas

In encountering any technical subject matter for the first time, you must, if you are to be able to understand it, learn the vocabulary of the subject matter. Think of all of the technical terms you had to learn in introductory psychology, arithmetic, algebra, economics, physics, chemistry, and so forth. There is also a technical vocabulary in elementary statistics that must be mastered.

In general, we try to express this vocabulary in the form of symbols and formulas. Symbols and formulas are, in reality, a form of shorthand — a simplified way of expressing something that would otherwise have to be written as a word, a phrase, or a sentence. Some of the symbols we shall use stand for quantities or numbers, and others stand for operations to be performed. Commonly used operational symbols are already familiar to you. They are $+$, $-$, $\times$, and $\div$, standing for the operations of addition, subtraction, multiplication, and division, respectively. Instead of using $\div$ as a symbol for division, we shall ordinarily represent this operation by means of a bar. Thus

$$\frac{2}{4}$$

or 2/4 means that 2 is to be divided by 4. To indicate multiplication, we shall frequently use parentheses. Thus (2)(4) means that 2 and 4 are to be multiplied. In cases where the meaning is clear, we may drop the parentheses and simply write ab. This means to multiply a and b.

We shall also use the symbols $<$ and $>$. The symbol $<$ means "is less than," and the symbol $>$ means "is greater than." Thus we can indicate that 2 is less than 5 by writing "$2 < 5$," and that 10 is greater than 7 by writing "$10 > 7$." If we write "$k \geq 5$," this is read "k is greater than or equal to 5," and "$k \leq 5$" is read "k is less than or equal to 5." Another symbol we shall use is an equal sign with a slash drawn through it, or $\neq$. This symbol means "not equal to." Thus, "$k \neq 3$" is read "k is not equal to 3."

A formula or equation always has one or more terms or symbols at the left and these are followed by an "$=$" sign. To the right of the $=$ sign there are also always one or more terms or symbols. As an example, which we shall discuss in detail later, we have

$$\bar{X} = \frac{\Sigma X}{n}$$

This formula states that $\bar{X}$ (a symbol for the arithmetic mean) is equal to the sum (the symbol Σ means to sum) of the values of X (a symbol for a variable), divided by n (a symbol for the number of values of X). Thus, expressed in words, this formula reads: "The arithmetic mean is equal to the sum of the values of X, divided by the number of values of X."

In essence, a formula that involves symbols enables us to say a lot with little wasted effort. A formula, in other words, is simply an abbreviated way of saying something.

1.4 Use of Tables

In the back of the book you will find a number of statistical tables that you will have occasion to refer to constantly. It is important that you know how to use these tables accurately. Each one will be explained in detail at the time it is first introduced in the text. One of these tables is a table of squares, square roots, and reciprocals of numbers from 1 to 1000. This table will enable you to obtain square roots easily and will also give you the squares of numbers so that you may avoid unnecessary multiplication. As a matter of fact, you will find it useful if you memorize the squares of numbers from 1 to 25. This is not a very great chore, because you probably already know at least half of these squares.

The reciprocal of a number is 1 divided by that number. You will recall that multiplication of one number by the reciprocal of a second number is the same as dividing the first number by the second. If you want to divide one number a by another number N, in some cases, you may find it easier to multiply the number a by the reciprocal of N, rather than actually dividing a by N.

1.5 Statistical Terms and Statements

1.5.1 AVERAGES

In our daily conversation, we often use the term *average*. We say, "John is better than average," when someone questions us about his golfing ability, or, "Mary is an average student and slightly below average in height." Some of our college courses we say we like "better than average." Although we may not have defined the term in our own thinking as precisely as a statistician would, we have some general understanding of the concept of "average." We may be vaguely aware that our statements concerning averages are based on a series of observations or measurements and that each of these observations, taken singly, may not be the same as the average we have in mind. We perhaps have some scale in mind when we refer to John's ability as a golfer or Mary's height and our average represents some middle position or value. The statement that "John is better than average" or that "Mary is slightly below average" indicates that we do not think of them in this middle position.

We can find statements similar to these in books about psychology and education. There we can find statements about the average intelligence quotient (IQ) of a group of children, the average score made by students on a reading test, the average number of errors made by rats in learning a maze, the average number of students per teacher in a given school, the average number of hours devoted to noninstructional activities by teachers in the same school, and so on. These statements about averages, however, would be expressed more precisely than the statements we make about averages in our daily conversation. In technical books and journals, averages are expressed in numbers, and the manner in which the numbers are obtained involves an elementary application of statistical methods.

1.5.2 VARIABILITY

A teacher recognizes the fact that, in her classes, not all students do equally well. On a test given to one class, some students may have very high scores and others may have very low scores. The same test given to another class may result in scores that are all very similar to one another. We would say that the students in the first class are more variable in their performance than those in the second. The psychologist finds that some rats learn to run a maze with very few trials and other rats require a large number of trials. Rats vary in their maze-learning abilities.

We constantly experience *variability* in daily life. We note variability in the teaching effectiveness of different teachers. We observe variability in the temperature from hour to hour, from day to day, and from month to month. We easily recognize the fact that some individuals are quite tall and others quite short. We find that some books are relatively easy to read, and others are difficult. Textbooks also differ or vary in their readability.

Statements about variability, like statements about averages, can be expressed precisely in the form of numbers. To measure the variability, sometimes called the dispersion, of a set of measurements also involves an application of elementary statistics.

1.5.3 RELATIONSHIPS

We make frequent reference to *relationships* in daily life, although these statements, like those we make about averages and variability, are not expressed precisely. We note that the amount of rainfall appears to be related to the season of the year, or that an individual's opinion on desegregation may be related to the section of the country in which he lives. We may say that John's golf game will improve with practice. In this instance, we indicate that we believe there is some relationship between amount of practice and performance on the golf course. The teacher observes that the scores students make on the first examination in a given course appear to be related to the scores they make on subsequent examinations.

If you will examine the textbook you used in introductory psychology, you will find many statements about relationships. These statements may refer to the relationship between grades in college and scores on an academic aptitude test or to the relationship between the IQ's of identical twins or to the relationship between any other two variables that have been investigated by research workers. Again, these statements are expressed precisely by means of numbers, and the method by which the numbers are obtained involves the application of elementary statistical methods.

1.6 Functions of Statistical Methods

1.6.1 PRECISE DESCRIPTION

One of the chief functions of statistical methods, as we have indicated, is to enable us to make precise statements about averages, variability, and relationships. With statistical methods, we do not have to rely on vague impressions about these things; we can describe them accurately.

1.6.2 PREDICTION

Suppose that we had studied a group of workmen operating a particular machine, and that we had then constructed a test of some sort that we believed to be related to performance on the machine itself. Giving the test to a large number of workers, we then find, by statistical methods, that the score on the test is related to performance on the machine. Could we then predict from the scores of a *new* group of workmen how well they would probably perform on the machine in question? If we find that there is a relationship between scores on a scholastic aptitude test and grades in college for one group of college students, can we use this information to predict the grades that another entering class of freshmen will make, knowing only their scholastic aptitude test scores, prior to their actually taking any college courses?

Statistical methods are also used in making predictions. Having made predictions, statistical methods then enable us to evaluate the accuracy and efficiency of the predictions. The problem of making predictions and evaluating predictions is also an important function of statistical methods.

1.6.3 STATISTICAL INFERENCE

Having made a series of observations and having described these particular observations in terms, let us say, of an average and a measure of variability, statistical methods, under certain conditions, enable us to take another step. Often, for example, our interest is not so much in the particular set of observations we have made, but rather in the larger supply or potential set of observations we could make, but have not. Suppose, for example, there are 12,000 students in attendance at a given school.

We have given a test to 400 of the students and for this group we have obtained a measure of their average performance and of the variability of their scores on the test. We call the particular 400 observations we have made a *sample*. The complete set of 12,000 potential observations we could make we call a *population*. If we obtain a descriptive measure for the sample, such as an average or a measure of variability, we call this measure a *statistic*. The corresponding measure for the population is called a *parameter*.

An important function of statistical methods is the problem of estimating population parameters from sample statistics, and the function is referred to as one of estimation. The process by which we arrive at such estimations is known as *statistical inference*. By this process we can generalize, under specified conditions, from knowledge of a characteristic of a sample to a characteristic of a population. In the case described above, for example, let us assume that the sample is representative of the population. We may wish to use our knowledge of the average and variability of the 400 sample observations as a basis for generalizing about the corresponding measures or parameters for the population of 12,000 observations. Because such generalizations are always subject to error, it is important to have some measure of the degree of confidence we have in making the generalization. The manner in which we do this is to specify limits within which we are relatively confident that a parameter falls. These limits are called *confidence limits*. The degree of confidence associated with these limits is called a *confidence coefficient*.

Another problem in statistical inference concerns whether or not it is reasonable to conclude that two or more averages, or other measures, differ significantly. In experimental work, for example, we may measure the performance of two groups of subjects, each under a different experimental condition or treatment. If the average performance of the two groups of subjects under the two experimental conditions is not the same, this may be the result of chance factors, rather than any difference in the experimental conditions themselves. Can we conclude that the difference in the average performance under the two experimental conditions is of such magnitude that it cannot reasonably be attributed to chance? If so, we say that the difference is significant. Statistical inference provides us with tests that can be used in judging the significance of differences. These tests are appropriately called *tests of significance*.

REVIEW OF TERMS

At the end of each chapter, we shall list the technical terms or concepts that have been introduced in the chapter. You should be able to give a verbal definition of each of these concepts or terms. Some of the terms used in this chapter have not been defined very precisely. We shall provide more precise definitions in later chapters.

average	population	statistical inference
confidence coefficient	relationship	test of significance
confidence limits	sample	variability
parameter	statistic	

REVIEW OF SYMBOLS

At the end of each chapter, we shall also list the new symbols that have been used in that chapter. Sometimes we shall include in this list symbols that have been used in earlier discussions. You should pay particular attention to these symbols and be prepared to define in words the meaning of each one.

Σ	X	$\bar{X}$
n	$\leq$	$\geq$
$<$	$>$	$\neq$

QUESTIONS AND PROBLEMS

1.1   Give the reciprocal, to four decimal places, of each of the numbers listed below. Use Table II in the Appendix.

(a) 12   (c)   14   (e) 15   (g) 10
(b) 25   (d) 146   (f)   4   (h) 75

1.2   Using Table II in the Appendix, find the reciprocals to four decimal places of:

(a) 8   (c) 0.08
(b) 0.8   (d) 0.008

1.3   Consider only the set of numbers 1, 2, 3, 4, 5, 6, 7, 8, and 9. We let k be any one of these numbers. Then what can you say about the possible values of k, if each of the following is true?

(a) $k < 3$ (d) $k > 6$
(b) $k \leq 4$ (e) $3 < k < 6$
(c) $k \geq 5$ (f) $k < 5$

1.4   Use Table II in the Appendix to find the square roots of each of the numbers listed below:

(a)   48,841   (e)   3.4225
(b) 137,641   (f)   5.1076
(c) 156,025   (g) 14.1376
(d) 110,889   (h) 24.9001

1.5   Use Table II in the Appendix to find the square roots of each of the following numbers:

(a) 220   (d) 4096
(b) 2.20   (e) 40.96
(c) 0.0220   (f) 0.4096

(g) 488.41
(h) 4.8841
(i) 0.048841
(j) 1376.41
(k) 13.7641

(l) 533.61
(m) 5.3361
(n) 1568.16
(o) 15.6816
(p) 0.156816

1.6 Why does a psychologist or a teacher need to have some understanding of elementary statistical methods?

1.7 What is the essential difference between a statistic and a parameter?

1.8 What is the essential difference between a sample and a population?

1.9 Give an example, in your field of interest, in which you would expect to find high values of one variable associated with high values of another variable and low values with low values.

1.10 Give an example, in your field of interest, in which you would expect to find high values of one variable associated with low values of another variable and low values with high values.

1.11 Would you expect high school freshmen to be more variable in their IQ's than college freshmen? Explain why.

1.12 Mary is 6 feet tall. Would you describe her as being above average, average, or below average in height? Explain why.

1.13 George weighs 120 pounds. Is it accurate to describe him as being below average in weight? Explain why. How would you describe him with respect to weight if he were ten years old?

1.14 All third-grade children in the public schools of the city of Seattle are given a standardized reading test. For each of the children an IQ is also available on a standardized test. Would you expect the reading test scores of the children to be related to their IQ's? In what way?

2

Variables and Scales

2.1 Introduction

The fundamental importance of observing things is recognized by all sciences. Those things that scientists observe are frequently called variates, or *variables*. For example, if the thing we are observing is the intelligence quotient (IQ) of children, the variable is the IQ. Any particular observation is a *value of the variable*. In our example, an observation of the particular IQ associated with a particular child is a value of the variable. The value of the variable indicates the *class* to which an observation is assigned. If the value of the variable is 100, we consider that observation and all others with a value of 100 as belonging to the same class. Obviously, if something is to be designated as a variable, we must have at least two possible classes of observations. These classes must also be *mutually exclusive*; that is, any particular observation can be assigned to only one of the available classes. By a variable, therefore, we mean anything that we can observe in such a way that each observation can be classified into one of any number of mutually exclusive classes.

2.2 Psychological Variables

In the behavioral sciences, one general class of variables we are concerned with is that relating to the behavior of organisms. By behavior, we mean any action of an organism. In some cases the behavior we observe may be a relatively simple response and in other cases a quite complex pattern of responses. The bar-pressing behavior of a rat in a Skinner box, for example, would be considered a relatively simple variable compared with

11

the aggressive behavior of a client in therapy or the withdrawal reactions of a child in a nursery school. The general class of things we observe relating to behavior we call *behavioral variables.*

Behavior, however, is not the only thing that a behavioral scientist observes. We have referred to the behavior of a rat *in* a Skinner box, the behavior of a client *in* therapy, the actions of a child *in* a nursery school. Behavior always occurs in a particular setting or environment and with respect to certain conditions of stimulation present in the environment. Our interest in the environment is based on the assumption that behavior in a particular setting is, in some degree, a response to that setting and the conditions of stimulation that it provides. The general class of things we observe that relate to the environment, situation, or conditions of stimulation, we designate as *stimulus variables.*

We observe that still another general class of things relates to properties or characteristics of organisms. We shall call this general class of variables *organismic variables.* For example, the thing we observe may be the color of hair associated with each of a number of individuals. Hair color would be an organismic variable and the specific hair color of a given individual would be an observation, or value, of the variable. Similarly, we might observe the heights and weights of a group of individuals. In this instance, also, we would refer to height and weight as organismic variables.

As a matter of convenience, we frequently make use of *response-inferred* organismic variables. IQ's, for example, are based on observations of the behavior or responses of individuals in a standardized testing situation. We may choose in a particular research problem, however, to regard the IQ as something that is associated with the organism and thus refer to it as an organismic variable, even though the IQ is based on observations of behavior. As another example, in a given experiment, we may select one group of subjects with high scores on the Taylor Manifest Anxiety Scale and another group with low scores on the same scale. The performance or behavior of the two groups may then be studied under certain experimental or stimulus conditions. We may find it convenient to refer to the first group as the "anxious" group and the second as the "nonanxious" group—that is, to treat anxiety, in this instance, as an organismic variable rather than as a response variable.

2.3 Scales of Measurement

2.3.1 NOMINAL SCALES

Suppose that the thing that we observe is the sex of each of a large number of students. Only two values of this variable are possible: we designate each student as a male or as a female. The basic data thus consist of the number of observations in each of the two classes, male and female. When we have ordered variables, which we shall discuss later, the various classes of the observations can be logically arranged, say, from left to right, so that we can say that any class falling to the left of a given class has less of the variable and any class falling to the right of a given class has more of the variable. Note that, in the case under discussion, the values of the variable are in no way ordered. If any observation is assigned the value "male," this does not indicate that it has a greater or lesser degree of the variable we have called sex than an observation assigned the value "female."

Observations of unordered variables, such as sex, are one of the most primitive forms of measurement and are described as constituting a *nominal scale*. In nominal scales, numbers may be substituted for the names of the various classes of the variable. The numbers serve only to identify the classes and do not indicate anything about the classes other than that they are different. The difference between the classes is one of *kind* and not degree. Thus we might identify all of the observations labeled "male" with the number 0 and all of the observations labeled "female" with the number 1. Because the numbers serve only to identify the classes, it would not matter if the males were identified by the number 1 and the females by the number 0. Other examples of variables in which observations constitute a nominal scale would be individuals classified according to psychiatric diagnosis, objects classified according to color, or citizens of the United States classified according to state of residence.

The data resulting from nominal scales are often referred to as *categorical data*, *frequency data*, *attribute data*, or *enumeration data*. The basic thing we have to deal with in the case of such data is a frequency, or the number of observations in each of the classes or categories.

2.3.2 ORDINAL SCALES

In some cases the observations we make may be ordered in such a way that we can say that one observation represents more of a given variable than another observation. For example, we might take five individuals and line them up in a row. By comparing the individuals to one another, we may be able to arrange them in order from what we judge to be the tallest to the shortest. If we identify the individuals by assigning the number 1 to the tallest, 2 to the next, 3 to the next, 4 to the next, and 5 to the shortest, the observations would be described as constituting an *ordinal scale*. The numbers used in identifying the observations are called *ranks*. Ranks tell us something about the degree of the variable within the set of observations at hand. The observation with Rank 1, for example, we believe represents a greater degree of the variable than any of the other observations in the set of five observations.

It is important to note that, although we believe ranks tell us whether one observation represents more or less of the variable of interest than another, they do not tell us *how much* more or less. For example, we know nothing about the exact difference in degree between any two ranks. The observation assigned Rank 1 may be, in measured units, 74 inches, that assigned Rank 2 may be 72 inches, and that assigned Rank 3 may be 66 inches. From the ranks alone, we know only that we believe the observation with Rank 1 is greater than that with Rank 2, and that with Rank 2 is greater than that with Rank 3, but we do not know how much greater.

2.3.3 INTERVAL SCALES

When the numbers used to identify observations represent not only an ordering of the observations but also convey meaningful information with respect to the distance or degree of difference between all observations, the observations are said to constitute an *interval scale*. Thus, if we have an interval scale, the numbers 7, 5, and 2, which may be the values of three different observations, tell us that 7 is 2 units greater than 5 and that 5 is 3 units greater than 2; that is, the difference between 7 and 5 is 2 units, and the difference between 5 and 2 is 3 units. The difference between 7 and 2 is the sum of the difference between 7 and 5 and the difference between 5 and 2, or $2 + 3 = 5$ units.

The ratio between the two intervals, $7 - 5 = 2$ and $5 - 2 = 3$, is 2/3. With an interval scale, we can add any given number to each of the values we have or subtract any given number from each of the values and still maintain the essential properties of the scale. The reason for this is that the *origin*, the value we choose to call zero, is arbitrary. Subtracting 2 from each of the three values, 7, 5, and 2, we then have 5, 3, and 0, and we have arbitrarily moved the origin to the value previously assigned 2. The ratio between the two new intervals, $5 - 3 = 2$ and $3 - 0 = 3$, is still 2/3.

The unit of measurement for interval scales is also arbitrary. If we consider the unit of measurement for the three values, 7, 5, and 2, to be 1, then we can double the unit of measurement by multiplying each value by 2 to obtain 14, 10, and 4 as the new values. Again we note that the ratio between the new intervals, $14 - 10 = 4$ and $10 - 4 = 6$, is 4/6 or 2/3, the same ratio we had before.

By adding or subtracting a given number from each value of a variable and by multiplying or dividing each value of the variable by any given number, other than zero, we obtain what is called a *linear transformation*[1] of the values on an interval scale. The essential properties of an interval scale remain unchanged by any linear transformation. Adding a specified

---

[1] A linear transformation Y of a variable X is any transformation in which a plot of $Y = f(X)$ against X in arithmetic coordinates is a straight line. Thus $Y = 2X$, $Y = X - 2$, or $Y = 3 + 2X$ are examples of a linear transformation of X because if the Y values are plotted against the corresponding X values in arithmetic coordinates, the plotted points, in each case, will fall on a straight line.

number to or subtracting a specified number from each value merely shifts the origin of the scale. Multiplication or division of each value by a specified number, other than zero, merely changes the unit of measurement. With interval scales, ratios between any two intervals remain invariant as a result of any linear transformation of the scale.

It is not, however, appropriate to take ratios between any two given *values* on an interval scale. We cannot, for example, with an interval scale, say that a value of 50 represents twice as much of the variable as a value of 25. Suppose, for example, we assume that scores on a test provide measures of a variable that we choose to call "vocabulary." We have three values of the variable, let us say, corresponding to scores of 25, 50, and 75, obtained by three different subjects. We cannot conclude that the score of 50 represents *twice the degree of the variable* as the score of 25 nor that a score of 75 represents three times as great a vocabulary as one of 25. To see why this is so, assume that we add 25 very easy items to the test so that no one of the three subjects fails to give the correct response to them. The three previous values of the variable would now become, with the new test, 50, 75, and 100, respectively. If we falsely believed that we could form ratios between values on an interval scale, we would now falsely assume that the person with a score of 100 had twice as great a vocabulary as the person with a score of 50, whereas the previous corresponding scores of 75 and 25 would have falsely led us to believe that the one person had *three* times as great a vocabulary as the other.

2.3.4 RATIO SCALES

The only scale on which ratios can be formed between any two given values of a variable is one in which we have an interval scale with an *absolute zero*. Scales of this kind are called *ratio scales*. Length, as measured in units of inches or feet, is a ratio scale, for the origin on this scale is an absolute zero corresponding to no length at all. Thus it is appropriate to state that an object that is 4 feet long is *twice* as long as an object that measures 2 feet.

In ratio scales, only the unit of measurement is arbitrary, the zero point being fixed. We cannot, therefore, add a given number to each value of the variable and still maintain the properties of the original ratio scale. For example, the ratio of the two values 4 and 2 feet is $4/2 = 2$ feet. If we add 5 feet to each of these two values, we have 9 and 7, and the ratio $9/7$ is not equal to 2, the value we obtained previously. To add a given number to each of the values or to subtract a given number from each of the values changes the nature of the scale.

Because the unit of measurement on a ratio scale is arbitrary, we can multiply or divide each of the values by a specified number, other than zero, without changing the properties of the scale. Thus, multiplying the values of 4 feet and 2 feet by 12, to change the unit of measurement from feet to inches, we have 48 and 24, and the ratio 48/24 equals 2, the same as before. In terms of inches, an object that is 48 inches long is still twice as long as one that is 24 inches long.

2.4 Discrete and Continuous Variables

We have described, briefly, four types of scales used in classifying observations of variables: the nominal scale, the ordinal scale, the interval scale, and the ratio scale. All of these scales constitute forms of measurement to the scientist, but in ordinary everyday language what we usually mean by measurements are values of a variable that is *continuous*.

Consider, for example, measurements of length. If we make observations of the heights of human subjects, we ordinarily record our observations in terms of the nearest inch. Although height is a continuous variable, height measured to the nearest inch is discrete. Some individuals may have, to the nearest inch, a measured height of 68 inches and others may have values of 67 or 69 inches. We say that these measurements are *discrete* because they can only fall at separated points on our scale in terms of our unit of measurement. There will be no values, for example, that fall between 67 and 68 inches or between 68 and 69 inches.

Obviously, however, we could have used a smaller unit of measurement, say 0.5 of an inch. The measured values would still fall at discrete points on this scale at, assume, the values of 60.0, 60.5, 61.0, 61.5, and so on. If we measured in terms of a still smaller unit, say 0.01 of an inch, we would have possible values of 60.00, 60.01, 60.02, and so on. If we take a still smaller unit, 0.001 of an inch, we would have possible values of 60.000, 60.001, 60.002, and so on. We note that no matter how small we make the unit of measurement it can always, in theory at least, be made smaller. Between any two points or values on the scale, no matter how small the unit of measurement, we can always think of another value that would fall somewhere between these two values. Measurements on a scale such as this would correspond to measurements of a *continuous variable*, even though the observed measurements are themselves discrete.

Measurements of continuous variables are always *approximations* of true values and are never exact. This is true of all continuous variables. Time may be measured in terms of years, months, weeks, days, hours, minutes, seconds, milliseconds, and so on, each succeeding unit being more precise than the one before; even milliseconds, however, are not exact values but only approximate. We have the interesting situation in which, although a variable is continuous, the recorded values of the variable are always, in practice, discrete.

Because of the approximate nature of measurements of continuous variables, we customarily regard height reported in terms of inches as representing an *interval*. For example, a height of 72 inches would be regarded as representing an interval ranging from 71.5 inches up to 72.5 inches, that is, half a unit of measurement below and half a unit above the value reported. If the unit of measure is 0.1 of an inch, then a value of 71.1 inches would be regarded as representing an interval ranging from 71.05 up to 71.15, that is, half a unit below and half a unit above the reported value.

If heights are observed to the nearest 0.1 inch but reported only to the nearest inch, we say that we have *rounded* the number. In rounding to the nearest whole number, we drop the decimal fraction if it is less than 0.5 and report the whole number. If the decimal fraction is over 0.5, we raise the whole number by one. If the decimal fraction is exactly 0.5, we drop it if the whole number is an even number and report the whole number. If the whole number is an odd number and the decimal fraction is 0.5, then the odd number is raised by one. The examples shown in Table 2.1 should make the accepted practice clear.

Table **2.1** — Rounding Numbers

8.635	rounded to two decimal places becomes	8.64
8.625	rounded to two decimal places becomes	8.62
7.51	rounded to one decimal place becomes	7.5
7.55	rounded to one decimal place becomes	7.6
7.45	rounded to one decimal place becomes	7.4
7.49	rounded to one decimal place becomes	7.5
7.43	rounded to one decimal place becomes	7.4
5.9	rounded to the nearest whole number becomes	6
6.5	rounded to the nearest whole number becomes	6
7.5	rounded to the nearest whole number becomes	8
5.3	rounded to the nearest whole number becomes	5

2.5 Number of Observations of a Given Kind as a Variable

Suppose that 10 coins are tossed and that we count the number of heads that appear. We have an observation of a variable for which the possible values are 0, 1, 2, . . . , 10. This variable obviously provides measurements on a ratio scale, because we have an absolute zero consisting of no heads at all. If one toss results in 8 heads and another toss results in 4 heads, it is obviously accurate to describe the first toss as resulting in *twice* as many heads as the second toss. Similarly, if one toss results in 9 heads and another in 3 heads, the first toss can be said to represent *three* times as many heads as the second toss.

In many psychological experiments, a variable of interest is the number of responses of a given kind. For example, in a discrimination experiment subjects may be given a series of trials in which they are asked on each trial to make a given discrimination. If in a set of 20 trials we count the number of correct discriminations, then we have a variable that can take the possible values of 0, 1, 2, 3, . . . , 20 correct discriminations. This variable is also measured on a ratio scale and it is perfectly proper to state that a subject who makes 18 correct discriminations has made *three* times as many correct discriminations as a subject who makes 6 correct discriminations.

Although the number of responses of a given kind must always be a discrete variable, we shall, as a matter of convenience, regard these counts in the same manner as we regard a reported discrete value of a variable that is continuous. For example, we shall regard a number such as 8 correct discriminations or responses of a given kind as representing an interval ranging from 7.5 up to 8.5. The reason we take this position with respect to the values of a discrete variable will be explained in a later chapter.

Suppose that students registered in an introductory psychology course are given a final examination consisting of 100 objective test items. Each item on the test might be regarded as a trial and the response to the item may be classified as correct or incorrect. Scores on the test are simply the total number of correct responses given to the items. Now, if one student makes 80 correct responses, then he has surely made *twice* as many correct responses as a student who makes only 40 correct responses. There is nothing wrong with this statement as long as we regard the variable being measured as the *number of correct responses.* In this case, we do have a ratio scale with an absolute zero representing no correct responses. On the other hand, if we regard the variable being measured as knowledge of the content of the introductory course in psychology, then we would most likely be in error in concluding that the student with a score of 80 has twice the knowledge that a student with a score of 40 has. If we regard scores on tests as measuring some variable other than simply the number of correct responses to the items in the test, we have at best an interval scale with an arbitrary origin and not a ratio scale with an absolute zero.

Regardless of whether we choose to view scores on tests as simply the number of correct responses or as a measure of some other underlying variable (such as knowledge, aptitude, or achievement of a given kind), the scores will be discrete. We shall also assume that a score on a test of, say, 50 represents an interval ranging from 49.5 up to 50.5. There are good reasons, as we shall see later, for viewing all recorded measurements as representing intervals, regardless of whether the variable being measured is in fact discrete or continuous.

2.6 Research Problems in Psychology and Education

Research consists essentially of stating questions or problems clearly, making observations that are relevant to the problem, analyzing and describing the observations, and interpreting the results of the analyses as they bear on the particular question or problem. Any question or problem for which it is possible to make relevant observations that will assist in answering the question or clarifying the problem can be made the basis of a research study.

The following questions illustrate something of the scope of research activities in psychology and education. Will children, on the average, work harder when they are praised than when they are criticized? Is there any relationship between grades earned in college and scores on a personality inventory? Will scores on a personality scale predict grades any better than scores on a scholastic aptitude test alone will? Does frustration result, on the average, in aggressive or regressive behavior or both? Will one method of teaching mathematics result in greater average achievement on the part of students than another method? Are boys more variable than girls in their performance on an achievement test? Do students, on the average, learn just as much from straight lectures as they do from lectures combined with quiz sections? In terms of average achievement, are small classes to be preferred to large classes? Are individuals who are honest in one situation likely to be honest in other situations? Do we tend to repress experiences that are unpleasant? Is there any real difference between the results obtained by counseling procedures used in "non-directive" therapy and "directive" therapy? To what extent can attitudes be changed as a result of viewing motion pictures? How can children's fears be eliminated most effectively?

Do white and black children differ in their average score on tests of intelligence? If so, how can we account for this difference? To what extent can children's intelligence test scores be modified by changes in the environment? Is it true that intelligence test scores of identical twins tend to be more similar than those of fraternal twins? Are personality traits related to measures of intelligence? Are boys more variable than girls in their scores on intelligence tests?

Are the attitudes we have toward various political concepts related to the meaning these concepts have for us? Will students learn just as much from a lecture presented over TV as they will from the same lecture presented in the classroom? Will individuals who are said to have a high degree of anxiety respond differently to stress than individuals with a low degree of anxiety? Are psychological tests useful in predicting recovery from mental illness? Can psychological tests be used to predict which individuals will be successful as airline pilots and which will not? How can we account for the migratory behavior of birds? How can we account for the fact that a pigeon released many miles away from its homing cage can find its way back to its cage? What is the most effective way of breaking a habit such as smoking? Why do some individuals become mentally ill and others do not?

The posing of a question, similar to those just listed, is the first step in research. Questions, when properly stated, become hypotheses that can then be subjected to empirical test. Once a question has been formulated, the next step is planning the research that we believe will provide a basis for answering the question. This consists of determining the relevance of various potential sets of observations that might be made, to the variables in which we are interested, the manner in which the observations will be made, and the methods by which they will be analyzed. The third step is actually carrying out the research and the analysis. The final steps are interpreting the analyzed data and seeing that the results are then made available to others.

In the questions raised above, note some of the problems involved. For example, in the first question, what do we mean by "work"? We must have in mind a response variable, but what observations shall we consider relevant to this variable? Do we mean work in a physical sense? By work "harder" do we mean "more persistently"? The meaning that we have attached to this variable will not be clear until we can *specify the observations* we are to make and the nature of the scale that we assume the observations comprise. With respect to "criticism" and "praise," we perhaps have in mind a condition of stimulation, but how shall we administer the "criticism" and "praise"? Again, this variable is not and will not be clearly defined for ourselves and others until the conditions of stimulation are specified.

The second question, merely by the form in which it is put, more clearly specifies the nature of the research problem. One variable will consist of grades earned in college and the other variables will consist of scores on a personality inventory. But do we mean all grades earned in college? Do we include grades in courses in physical education? Over what period of college attendance will the grades be based? Shall we demand at least one year of grades? Will grades in all courses be considered equivalent? Do we include both males and females in the study? Graduate students? And which of the many available personality inventories shall we use in the research?

REVIEW OF TERMS

absolute zero   continuous variable
arbitrary origin   discrete variable
behavioral variable   interval scale

linear transformation of a scale   ratio scale
nominal scale   response-inferred organismic variable
ordinal scale   stimulus variable
organismic variable   variable
ranks

QUESTIONS AND PROBLEMS

2.1   Although it was stated that all recorded measurements are discrete, some variables are continuous. Indicate which of the following variables you would regard as continuous and explain why:

(a) gains in weight of infants from age 6 months to age 12 months
(b) reaction times
(c) speed of running a maze
(d) number of students in each class at a university
(e) number of children per married couple
(f) annual income of professors at a university
(g) time spent in preparing an assignment
(h) number of correct responses on a test
(i) number of trials required to learn a list of words
(j) number of times that a coin falls heads in ten tosses

2.2   Which of the variables listed in 2.1 could be described as being measured on a ratio scale?

2.3   Round the following numbers to two decimal places:

(a) 28.34501   (f) 133.4550
(b) 2.68400   (g) 18.4650
(c) 32.35500   (h) 23.7861
(d) 1.84732   (i) 22.6850
(e) 16.36491   (j) 13.7950

2.4   Patients in a mental hospital are classified as: paranoids, schizophrenics, manics, hysterics, etc. What kind of a scale is involved?

2.5   On December 22 at 3:00 P.M. the temperature in Anchorage was 20 degrees below zero, whereas at the same hour the temperature in Minneapolis was 10 degrees below zero. Was it twice as cold in Anchorage as in Minneapolis? Explain your answer.

2.6   Scientists engaged in research on cigarette smoking and lung cancer sometimes classify individuals as nonsmokers, light smokers, and heavy smokers. What are some of the problems involved with this scheme of classification? What kind of scale do you think is involved in this classification?

2.7   How does an ordinal scale differ from an interval scale?

2.8   Students enrolled at a university are classified as freshmen, sophomores, juniors, seniors, and graduate students. What kind of a scale do you think is involved in this classification?

2.9   It is known that Jim smokes 20 cigarettes every day and that Joe smokes 10 cigarettes every day. Comment on the following two statements: (1) Jim smokes twice as many cigarettes each day as Joe; (2) Jim smokes twice as much as Joe. Would either statement need to be qualified if it were known that Jim smokes regular cigarettes whereas Joe smokes king-size cigarettes? If Jim smokes Vantage cigarettes and Joe smokes Lucky Strike regulars, would either statement need to be qualified? Would it make any difference if Jim takes only two puffs on each cigarette whereas Joe takes ten puffs on each cigarette?

2.10   Mr. Smith's annual income is $30,000.00 and Mr. Johnson's annual income is $20,000.00. Is it correct to say that Mr. Smith's annual income is 1.5 times that of Mr. Johnson's?

2.11   Does a major in the air force "outrank" a captain? Does an associate professor in a department of psychology "outrank" an assistant professor?

2.12   Suppose you were told that Michael ranked tenth from the top in academic standing in his high school graduating class. In interpreting this statement would you need to know the number of students in his graduating class? Would it make any difference, for example, if his graduating class consisted of 500 students or only 15 students?

2.13   Explain why all observed or recorded measurements of continuous variables must, of necessity, be approximate and discrete.

2.14   John has a Stanford-Binet IQ of 150 and Sam's rating is 75. A student states that John is twice as intelligent as Sam. Do you agree with this statement? Explain why.

2.15   What is the essential difference between a nominal scale and the other forms of measurement described in the chapter?

2.16   How does a ratio scale differ from an interval scale?

2.17   Would the rank order of a given set of observations be changed (a) if each rank were multiplied by 5, or (b) if 10 were added to each rank? Explain why.

2.18   Read an article in a journal devoted to research in your subject area. Give a reference to the article in the same manner in which references in the journal are cited. (a) What question or questions was the research attempting to answer? (b) What variables were involved in the research? (c) What observations served to define the variables?

2.19   Discuss some possible ways in which you might attempt to measure individual differences in each of the following variables and also the nature of the scale which the measurements would comprise:

(a) aggressiveness
(b) interest in reading science fiction
(c) teaching effectiveness
(d) neuroticism
(e) skill in driving an automobile

(f) typing ability
(g) creativity
(h) achievement in a course in statistical methods
 (i) motivation to achieve good grades in college
 (j) attitude toward the United Nations
(k) knowledge of current affairs
 (l) popularity
(m) skill in playing table tennis
(n) visual accuity

3

Frequency Distributions

3.1 Introduction

Once we have made a set of observations, what we do with them depends on the questions we raised prior to making the observations and how we expect the observations to assist us in answering these questions. We emphasize again that we ordinarily undertake the kind of systematic observation that characterizes research because we believe it will help us in answering questions of interest.

One question we might ask concerns the manner in which the observations are distributed among the various classes of a variable. For example, suppose the observations consist of the IQ's obtained from a sample of 1000 children. Although we have 1000 observations, we shall not have 1000 different numerical values. We might wish to know whether we have more observations with a value of 130 than with a value of 70. Or we might wish to know which value occurs most frequently in the sample of 1000. We are interested, in other words, in how the observations are distributed among the possible classes of the variable.

By counting, we may find that we have four observations such that the value of each is 89, six with the value 90, ten with the value 91, and so on. As we have pointed out earlier, all IQ's that are numerically the same are considered as belonging to the same class. The number of observations in a given class is called the *frequency* of that class. Thus the class 89 would have a frequency of 4, 90 a frequency of 6, 91 a frequency of 10, and so on. When we have recorded the classes of a variable and the frequency for each class, we have a *frequency distribution*. A frequency distribution shows how the observations are distributed among the various classes of a variable.

When we are concerned with the distribution of observations of a single variable, the distribution is described as *univariate*. When the observations consist of pairs of values, one for each of two variables, we describe the distribution as *bivariate*. In this chapter, we shall illustrate tabular and graphic methods useful in describing univariate and bivariate distributions.

3.2 Univariate Distributions

Suppose that a variable of interest is response to an item in a test. If the item is a true-false item, with one of the answers keyed correct, any single value of the variable can take the form Right (R) or Wrong (W), depending on the response observed. Table 3.1 gives the results of 50 observations recorded a column at a time in the order in which the observations were made. The distribution of the observations is not readily apparent from the order in which they were originally recorded. The nature of the distribution is, however, clearly perceived in Table 3.2, where we show the number of observations falling in each of the two classes. We have reduced the original 50 observations to two numbers, the two frequencies. We use the letter f to indicate frequency in the table. These two numbers summarize very effectively, in the form of a frequency distribution, all of the observations we have made.

We can also portray the frequency distribution shown in Table 3.2 in the form of a bar chart or bar diagram, as shown in Figures 3.1 and 3.2. In the one case, Figure 3.1, we have let the heights of the bars correspond to the observed frequencies of 30 R's and 20 W's. In the second case, Figure 3.2, we have expressed these two frequencies as proportions by

Table **3.1** — Record of 50 Observations Classified as Right (R) or Wrong (W)

R	R	R	R	R	R	R	W	R	R
R	W	W	W	W	R	W	R	W	W
W	R	R	R	W	W	R	R	W	R
W	R	W	W	R	R	R	W	R	W
R	W	R	R	R	W	W	R	R	R

Table **3.2** — Frequency Distribution of 50 Observations Classified as Right or Wrong

(1) Response	(2) f	(3) p
Right	30	0.60
Wrong	20	0.40
Total	50	1.00

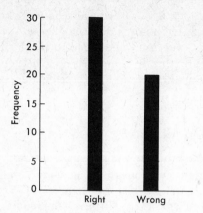

Figure 3.1 — Bar diagram showing the frequency distribution of 50 observations classified as Right or Wrong.

dividing each by the total number of observations. Thus we have $30/50 = 0.60$ of the observations with a value of R, and $20/50 = 0.40$ of the observations with a value of W. We use the letter p to indicate proportions in the table.

Table 3.3 shows the frequency distribution of responses to an item in an attitude scale. The five categories or values of the variable consist of responses of Strongly Disagree (SD), Disagree (D), Undecided (U), Agree (A), and Strongly Agree (SA). The table gives both the frequency for each of these classes and the proportion of the total number of observations in each class. The proportions were obtained by dividing each of the frequencies by the total number of observations, 200. Figure 3.3 is a bar diagram that gives a graphic portrayal of the frequency distribution corresponding to the tabular presentation of the distribution in Table 3.3.

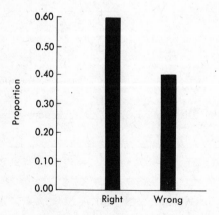

Figure 3.2 — Bar diagram showing the proportion of 50 observations classified as Right or Wrong.

Table **3.3** — Frequency Distribution of 200 Responses to an Item in an Attitude Scale

(1) Response	(2) f	(3) p
Strongly disagree	60	0.30
Disagree	80	0.40
Undecided	30	0.15
Agree	20	0.10
Strongly agree	10	0.05
Total	200	1.00

Table 3.4 gives the frequency distribution of scores obtained by 749 female college students on a personality scale that was designed to measure a variable called Need for Achievement. The efficiency with which the frequency distribution summarizes the 749 observations is obvious, in this instance, if we imagine trying to understand the nature of the distribution from the individually recorded 749 observations.

For each of the scores in Table 3.4, the unit of measurement is one scored response to an item. Each score can differ from the next highest or the next lowest score by only one unit. In an earlier discussion we said that we would regard scores on tests as intervals, so that, in the present instance, all scores of, say, 3 are regarded as occupying an interval ranging from 2.5 up to 3.5. We also assume that all scores classified in this interval can be best represented by the *midpoint* of the interval, or the value 3.0,

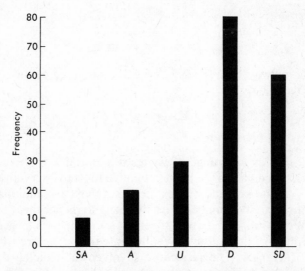

Figure **3.3** — Bar diagram showing the frequency distribution of 200 responses to an item in an attitude scale.

Table **3.4** — Frequency Distribution of Scores Obtained by 749 Female College Students on a Scale Designed to Measure Need for Achievement

(1) Score	(2) f
28	1
27	3
26	2
25	3
24	2
23	4
22	7
21	8
20	22
19	22
18	46
17	37
16	52
15	60
14	45
13	83
12	82
11	67
10	60
9	45
8	38
7	22
6	15
5	8
4	6
3	4
2	2
1	3

which falls midway between 2.5 and 3.5. This score continuum is illustrated in Figure 3.4, *A*.

3.3 Grouping Observations

In many instances, the range of values of a variable will exceed the 28 different recorded values we happen to have for the observations given in Table 3.4. As the range of values increases, we have additional classes and, with an excessive number, we may desire to reduce these to a smaller number. We do this by putting together in a single class a number of adjacent values of the variable. We illustrate the procedure with the data shown in Table 3.4. The procedure involved in reducing the original classes of a variable to a smaller number is sometimes referred to as *grouping the observations into broader classes.*

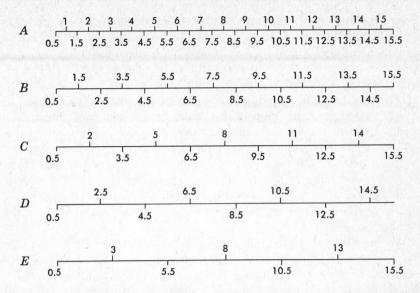

Figure **3.4** — Scales showing the midpoints and limits for different size intervals. Midpoints are indicated by the numbers above the line and limits by those below. A: Midpoints and limits with an interval of 1. B: Midpoints and limits with an interval of 2. C: Midpoints and limits with an interval of 3. D: Midpoints and limits with an interval of 4. E: Midpoints and limits with an interval of 5.

Suppose we combine the classes 1 and 2, 3 and 4, 5 and 6, and so on. This means we are willing to ignore the originally observed difference between the values of 1 and 2, for example, and to treat them as equivalent, and similarly for the other combined classes. The interval of classification for scores 1 and 2 now becomes 0.5 up to 2.5, and that for scores 3 and 4 becomes 2.5 up to 4.5, and so on, as shown in Figure 3.4, *B*. Again we assume that the best single value to represent all scores in a given interval is the midpoint of the interval which, in the case of the scores in the interval 0.5 up to 2.5, is 1.5. Similarly, the scores of 3 and 4 classified in the interval from 2.5 up to 4.5 are assumed to be best represented by the midpoint of this interval which is 3.5, and so on. We note that, on this scale, the values of the variable increase by 2 units, starting with 1.5, going to 3.5, and so on.

Figure 3.4, *C, D,* and *E,* illustrate the scales we would obtain if we combined the original values into class intervals of three units, four units, and five units, respectively.

Obviously, the broader we make the class intervals for any given set of observations, the smaller the number of new classes we will obtain. It is also true, however, that as we increase the width of the class intervals, the greater the degree to which we tend to ignore differences existing between the original values assigned to the new classes. If we desire to reduce the original observations into a smaller number of classes, what is the minimum number of new classes we should seek so as not to classify as equivalent too wide a range of values? No hard and fast rules are available. General practice would seem to indicate a minimum of 10 classes with, perhaps, 15 being a reasonably good average number of classes.

For any given distribution, the width of the interval to be used to obtain a good approximation of the desired number of class intervals is readily determined. We first find the range of the values of the variable. The *range* is the difference between the largest and the smallest value and is one measure of the variability of a set of observations. For the data of Table 3.4, the range is $28 - 1 = 27$. If we divide the range by the number of class intervals desired and round the result to the nearest integer, we will have the width of the interval to use to obtain approximately the number of desired classes.

We see that for the data of Table 3.4, the original observations range from 28 to 1. If we are to have approximately 15 classes, we should not use a class interval greater than 2. Table 3.5 shows the resulting frequency distribution of the original 749 observations with the new class interval of 2. The proportion of the total number of observations falling in each of the new intervals is also shown in column (3). The proportions were found by multiplying each of the frequencies in column (2) by the reciprocal of the total number of observations, that is, by $1/749 = 0.00133511$.

Table **3.5** — Frequency Distribution of 749 Scores with a Class Interval of 2

(1) Scores	(2) f	(3) p	(4) cf	(5) cp
27–28	4	0.005	749	1.001
25–26	5	0.007	745	0.996
23–24	6	0.008	740	0.989
21–22	15	0.020	734	0.981
19–20	44	0.059	719	0.961
17–18	83	0.111	675	0.902
15–16	112	0.150	592	0.791
13–14	128	0.171	480	0.641
11–12	149	0.199	352	0.470
9–10	105	0.140	203	0.271
7–8	60	0.080	98	0.131
5–6	23	0.031	38	0.051
3–4	10	0.013	15	0.020
1–2	5	0.007	5	0.007

In Figure 3.5, we show the bar diagram corresponding to the frequency distribution of Table 3.5, with the height of the bars representing the proportion of the total number of observations falling in each of the new class intervals. When we have an ordered variable, as we do with these

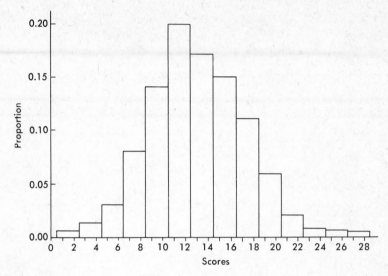

Figure **3.5** — Histogram showing the distribution of scores of 749 college females on a scale designed to measure Need for Achievement.

test scores, the bar diagram portraying a frequency distribution is also called a *histogram*. It obviously would also be possible to portray the distribution in terms of the actual frequencies in the classes instead of the proportions in the classes.

3.4 Cumulative Distributions

When values of a variable can be ordered, as in the case of test scores, we may also portray the frequency distribution in the form of a *cumulative distribution*. By a cumulative distribution we mean that we cumulate the frequencies as we move from the lowest to the highest interval, as shown in column (4) of Table 3.5. The successive entries in this column give the sum of all of the frequencies falling below the successive upper limits of the class intervals. For example, the entry of 98, corresponding to the class interval 7–8, is obtained by adding the frequencies of 5, 10, 23, and 60 to obtain 98, the number of observations that fall below the upper limit, 8.5, of the class interval 7–8.

The cumulative frequencies (*cf*), shown in column (4) of the table, are expressed as cumulative proportions (*cp*) in column (5). These values were obtained by cumulating the proportions in column (3) in the same manner in which we cumulated the frequencies. A graphic portrayal of the cumulative distribution is shown in Fig. 3.6. Note that the points plotted in the graph correspond to the recorded *upper limits* of the class intervals. This is because the cumulative proportions are matched with

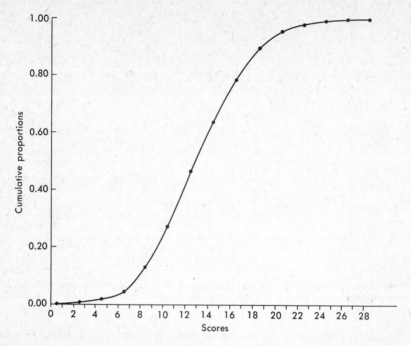

Figure **3.6** — Cumulative-proportion graph of the scores of 749 college females on a scale designed to measure Need for Achievement.

or correspond to the upper limits of the successive class intervals, rather than the midpoints of the class intervals. By drawing a smooth line through the successive values of the cumulative proportions, we obtain a graph that is called a *cumulative-proportion graph*.

3.5 The Shape or Form of a Distribution

The shape or form of frequency distributions observed in educational and psychological measurement varies. In Figure 3.7, we show the bar charts corresponding to the frequency distributions of four different sets of observations. For the first distribution, we note that the score of 6 occurs with the greatest frequency. In frequency distributions, the interval with the greatest frequency is called the *modal interval*, and the value of the observation that occurs most frequently is called the *mode*. For the first distribution, *A*, we observe that the frequencies of the two extreme intervals are relatively small. As we move in toward the middle interval from the two extremes, the frequencies increase until we reach the modal interval, which is in the center of the distribution. For frequency distributions of this kind, the cumulative-proportion graph resembles an S-shaped curve, as we see in Figure 3.8, *A*.

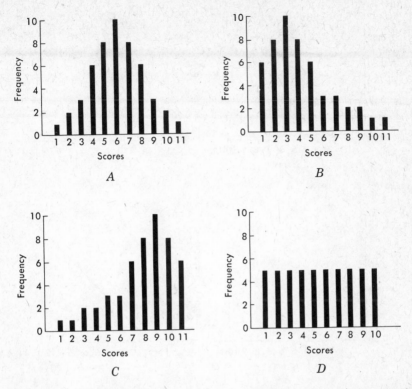

Figure **3.7** — Illustration of frequency distributions with different forms. A: A symmetrical distribution with a tailing off in both directions from the mean. B: A distribution with a tail to the right; a positive, or right-skewed, distribution. C: A distribution with a tail to the left; a negative, or left-skewed, distribution. D: A rectangular distribution.

The mode for the second distribution of Figure 3.7, B, is not at the center of the distribution but toward the left or low end of the scale. The frequencies, we observe, decrease or tail off as we move toward the higher end of the scale. For the third distribution, C, the situation is reversed, with the mode falling toward the high end of the scale and with a tailing off toward the low end of the scale. Both of these distributions are said to be *skewed*. The direction of the skewness is indicated by the tail of the distribution. When the tail is to the right, as in the case of the second distribution, B, the distribution is described as skewed to the right, or as positively skewed. When the tail is to the left, as in C, the distribution is described as skewed to the left, or as negatively skewed. Note the difference in the shape or form of the cumulative-proportion graphs of these two distributions, as shown in Figure 3.8, and also how the two graphs differ from that for the first distribution.

For the fourth distribution, Figure 3.7, D, the frequency of each interval is the same, and the distribution is described as *rectangular* or *uniform*.

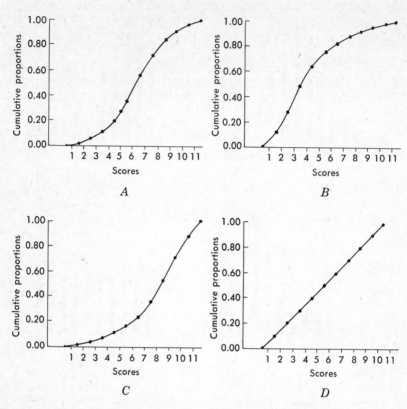

Figure **3.8** — Cumulative-proportion graphs for the frequency distributions of Figure 3.7. A: The cumulative-proportion graph for a symmetrical frequency distribution with a tailing off in both directions from the mean. B: The cumulative-proportion graph for a positive, or right-skewed, frequency distribution. C: The cumulative-proportion graph for a negative, or left-skewed, frequency distribution. D: The cumulative-proportion graph for a rectangular distribution.

The cumulative-proportion graph of a uniform distribution is a straight line, as shown in Figure 3.8, *D*.

3.6 Bivariate Distributions

Let us suppose that we have two variables of interest and that the observations of each variable are paired so that for each value of one variable we have a corresponding value of the second variable. In the simplest case, we might consider responses to two different items in a test, such that each response to each item is classified as R or W. To indicate whether the observation refers to a value of the first or second variable, we use the subscripts 1 and 2 to refer to the first and second variables, respectively, so that we have R_1 and W_1 and R_2 and W_2 as possible values of the two variables. It will be convenient if we think of a single observation as consisting of one of the possible pairs of values of these two variables. The possible pairs of values are, obviously, R_1R_2, R_1W_2, W_1R_2, and W_1W_2.

Ordinarily, when we have paired values of two variables, our interest centers in the possible association or relationship that may exist between the values of the two variables. Again, in order to perceive whether a relationship may or may not exist, we resort, for the time being, to tabular and graphical methods. In the case at hand, the basic data will consist of the frequency with which each of the possible pairs of values occurs. Table 3.6 shows the schematic representation for the frequencies of each of the possible pairs of values. Table 3.7 shows the corresponding observed frequencies for each of these pairs for a set of 100 observations.

Table **3.6** — Schematic Representation of Distribution of Responses to Two Items with Responses to Each Item Classified as Right or Wrong

Variable 1	Variable 2		Total
	Wrong	Right	
Right	R_1W_2	R_1R_2	R_1
Wrong	W_1W_2	W_1R_2	W_1
Total	W_2	R_2	

Table **3.7** — Distribution of 100 Responses to Two Items

Variable 1	Variable 2		Total
	Wrong	Right	
Right	10	50	60
Wrong	20	20	40
Total	30	70	100

We note that the *marginal* totals of 60 and 40 give the frequencies for R_1 and W_1, that is, the number of Right and Wrong responses to the first item. The marginal totals of 70 and 30 give the corresponding frequencies for R_2 and W_2. The cell entries of the table, 10, 50, 20, and 20, give the frequencies of the paired values of R_1W_2, R_1R_2, W_1W_2, and W_1R_2, respectively. As we shall show later, if the response to the first item is in no way related to or associated with the response to the second item, then the ratio of the cell entries R_1W_2/W_1W_2 should be equal to the ratio of the cell entries R_1R_2/W_1R_2. For the data of Table 3.7, these ratios are 10/20 and 50/20, and the fact that they are not equal indicates that there is some association between the two variables. If the cell frequencies were as shown in Table 3.8, the corresponding ratios would be 18/12 and 42/28, and these two ratios are equal. Thus, for the data of Table 3.8, there is no indication that the two variables are associated.

Table 3.8 — Theoretical Distribution of 100 Responses to Two Items which Are Not Associated but with the Same Marginals as Table 3.7

Variable 1	Variable 2		
	Wrong	Right	Total
Right	18	42	60
Wrong	12	28	40
Total	30	70	100

To consider another case, suppose that, as before, we have responses to two items in an attitude scale in which the possible values of both of the variables consist of the responses SA, A, U, D, and SD. Table 3.9 gives the frequency with which each of these values occurs for both of the variables. We note from the cells of the table that 25 different possible pairs of values or observations are possible. We observe also that there is an apparent tendency for a response to the first item to be associated with a similar response to the second item. For example, if we consider the ten observations classified as SA on the first item, we find that eight of these are classified as SA on the second item. From the table alone, however, we can get only some indication of the nature of the association between the two variables. In a later chapter, we shall describe a method for obtaining a quantitative measure of the degree of association between the two variables.

Table 3.9 — Distribution of 100 Responses to Two Items in an Attitude Scale

Item 1	Item 2					
	SD	D	U	A	SA	Total
SA				2	8	10
A			1	15	4	20
U		6	20	4		30
D	1	18	6			25
SD	10	4	1			15
Total	11	28	28	21	12	100

We now consider the case of measurements on an interval, ratio, or ordinal scale. First, we assume that we have scores on two psychological tests and that these scores represent values of the two variables we are interested in. Again, we are interested in determining whether or not there is any association or relationship between the two variables. Because the range of values on both variables is fairly large, we have first reduced the number of classes for one variable to 8 and for the other to 9.

For the data of Table 3.10, we find that there is a tendency for a high value of the first variable to be associated with a high value of the second variable. And, similarly, we find that there is a tendency for a low value of the first variable to be associated with a low value of the second variable.

Table **3.10** — Distribution of 32 Scores on Two Psychological Tests

Test 1	Test 2									
	100–104	105–109	110–114	115–119	120–124	125–129	130–134	135–139	140–144	Total
21–23						1	1	1		3
18–20				1						1
15–17					1	2				3
12–14		1	1				2	1	1	6
9–11				1			1			2
6–8		1		2		1	2	1		7
3–5			1		1		1			3
0–2	5		1			1				7
Total	5	2	3	4	2	5	7	3	1	32

When we have two variables such that high values of one variable are associated with high values of the second, and low values of the first are associated with low values of the second, we describe the relationship between the two variables as *positive*. On the other hand, if high values of one variable are associated with low values of the second, we say that the relationship between the two variables is *negative*.

The notion of positive and negative associations or relationships simply refers to the direction of the association between the two sets of values. A positive relationship means that high values of one variable are associated with high values of the second and that low values of the first are associated with low values of the second. Negative relationships, on the other hand, are those in which high values of the first variable are associated with low values of the second variable and low values of the first variable are associated with high values of the second variable. This description of relationships as positive or negative is meaningful only when the values of both variables can be ordered.

Consider still another kind of problem. We assume that we have rank orders available for each of a given set of objects on each of two different variables. Suppose, for example, we decide that ambiguity of Thematic Apperception Test (TAT) cards can be defined in two different ways. We provide a judge with these two definitions and ask him to rank order each of 10 TAT cards in terms of the definitions provided. We are interested in determining whether there is any relationship or association between the values assigned to the cards when judged in terms of the two definitions. Table 3.11 gives the ranks assigned to each of the ten cards by a given judge, with the rank of 1 indicating his judgment of greatest ambiguity and the rank of 10 the least ambiguity. To determine, graphically, whether there is any relationship between the two sets of ranks, we plot the corresponding values as shown in Figure 3.9.

Table 3.11 — Ranks Assigned to 10 TAT Cards under Two Different Definitions of Ambiguity

	Ranks Assigned	
Card	Definition 1 X	Definition 2 Y
A	8	10
B	7	9
C	10	8
D	9	7
E	3	6
F	4	5
G	6	4
H	5	3
I	2	2
J	1	1

This graph indicates immediately that there is a decided tendency for the ranks assigned to the cards under one definition of ambiguity to be associated with the ranks assigned to the same cards under the second definition by the judge. Each point in the graph corresponds to a given

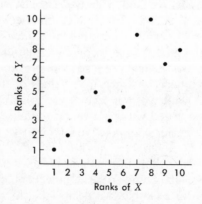

Figure 3.9 — Plot of the X and Y ranks given in Table 3.11.

pair of ranks. We can designate the rank assigned to a given card under the first definition as X and under the second definition as Y. Each point plotted in the graph thus corresponds to a pair of (X,Y) values. It is customary to represent the X values on the horizontal scale and the Y values on the vertical scale. Any given pair of values, consisting of the X and Y values, is designated by (X, Y) with the X value given first.

We consider one final case. Suppose that the values of X consist of the successive trials and the values of Y consist of the number of errors made by a given subject in learning a maze during the various trials. The basic data are shown in Table 3.12, and Figure 3.10 gives a plot of the number of errors against the number of trials. In this instance, it is obvious that as the number of trials increases the number of errors decreases, and we say that the two variables are negatively related.

Table **3.12** — Number of Errors Made on Each of 10 Trials in Learning a Maze

Trials X	Number of Errors Y
1	18
2	12
3	9
4	6
5	5
6	4
7	3
8	2
9	2
10	1

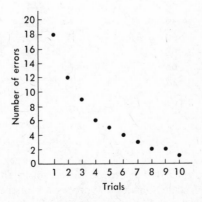

Figure **3.10** — Plot of errors against trials for the data of Table 3.12.

In this section we have described ways in which the relationship between the paired values of two variables can be portrayed in tabular or graphic form. We shall, in later chapters, show how to describe quantitatively the relationship between the paired values of two variables.

REVIEW OF TERMS

bivariate distribution	mode
class interval	modal interval
cumulative-proportion graph	negative relationship
frequency distribution	positive relationship
grouping observations into	proportion
broader classes	range
histogram	skewed distribution
limits of a class interval	uniform distribution
midpoint of a class interval	univariate distribution

REVIEW OF SYMBOLS

f p

cf cp

QUESTIONS AND PROBLEMS

3.1   The scores given below were obtained by 100 students on a vocabulary test. (a) What is the range of scores? (b) What is the lower limit of the score 44? (c) Make a frequency distribution, using a class interval of 5. Let the lowest or first class interval include 30, 31, 32, 33, and 34. What is the lower limit of the first class interval? (d) What is the midpoint of the class interval 60-64? (e) What is the modal class?

87	76	73	70	67	66	64	63	61	60
85	75	72	69	67	65	64	62	61	60
82	74	71	69	67	65	63	62	61	60
78	74	71	68	66	65	63	62	61	60
77	74	70	68	66	64	63	62	61	60

60	59	58	57	56	54	52	50	46	43
60	59	58	57	55	54	52	49	46	42
60	59	58	57	55	53	51	49	46	38
60	59	58	56	55	53	51	48	45	35
60	59	57	56	54	53	50	47	44	33

3.2   For the grouped data of 3.1, draw a histogram to portray the distribution with the vertical scale showing the proportion of the total number of observations in each class interval. Label both the horizontal and vertical scales. Put a title under the figure.

3.3 Using the grouped data of 3.1, draw a cumulative-proportion graph. Label both the horizontal and vertical scales and put a title under the figure.

3.4 A total of 80 students practiced with experimental materials until they had reached a certain criterion of learning. The students were then divided into eight groups of ten students each. A different group was tested for retention of the learned material on each of eight successive days. The Y values correspond to a measure of loss in retention; that is, the larger the Y value, the greater the loss in retention. (a) Make a graph showing the loss in retention with successive days. (b) Would you describe the relationship between these two variables as positive or negative? Explain your answer.

X Days	Y Average Loss in Retention
1	24.7
2	41.7
3	55.6
4	56.4
5	60.1
6	66.3
7	70.3
8	77.0

3.5 On a test it was found that of 200 students, 100 gave the Right and 100 gave the Wrong response to Item 1. On Item 2, 110 students gave the Right response and 90 gave the Wrong response. Of the 200 students, 60 gave the Right response to both items. (a) Set up a 2×2 table with the cell entries corresponding to the frequencies of each possible pair of values, that is, R_1R_2, R_1W_2, W_1R_2, and W_1W_2. (b) Is there any association between response to the first item and response to the second item? Explain your answer. (c) Assume that the marginal totals for the 2×2 table remain the same. If this is the case, then what would the cell entries in the table have to be if the two variables are not associated?

3.6 What are some of the considerations that should enter into a decision to group observations into broader classes?

3.7 Assume that we have six different frequency distributions and that for each distribution we have a different sized class interval. Below we give only the lowest class interval for each of the different frequency distributions. Give the limits and midpoints for each of the lowest class intervals listed.

(a) 20–23   (d) 0.10–0.20
(b) 9–12   (e) 2.5–3.5
(c) 120–124   (f) 0.85–0.95

3.8 Suppose that we have six different sets of observations with the range of values as indicated below. Assume that you want to group the observations in each set into broader classes and that you want to have approximately 15 class intervals for each distribution. What size class interval would you use for each distribution?

(a) Range = 62–32
(b) Range = 98–15
(c) Range = 73–13
(d) Range = 160–14
(e) Range = 88–45

3.9 Having decided upon the size class interval to use for each of the distributions of Problem 3.8, assume that you want the recorded lower limit of the first class interval to be a multiple of the size of the interval. For example, if we have a class interval of 3 and if the smallest value of a variable is 7, the recorded lower limit of the first interval would be 6, if the first interval is to include 7 and, at the same time, be a multiple of the size of the interval. (a) For each of the distributions in Problem 3.8, give the recorded lower and upper limits of the first class interval, if the lower recorded limit is to be a multiple of the size of the interval and if the first interval is to include the lowest observed value of the variable. (b) What are the midpoints of each of the lowest class intervals?

3.10 Of the four types of scales described in Chapter 2, for which is the range a sensible measure? Explain why.

3.11 What is the difference between a univariate and a bivariate distribution?

3.12 In order to describe a relationship between two variables as positive or negative, what must be true of the values of both of the variables?

3.13 Suppose that we have a sample of 1000 families and that we count for each family the number of children in the family. What would you predict about the shape or form of the frequency distribution? Explain why.

3.14 A vocabulary test is standardized on fifth-grade children and for these children the distribution of scores is similar in shape to the distribution shown in Figure 3.7, A. If the test is given to tenth-grade children, what would you predict about the shape or form of the frequency distribution of scores? Explain why.

3.15 Assume that we have responses of 100 students to an item in a standardized achievement test of arithmetic (X) and also to an item in a standardized vocabulary test (Y). The response to each item is scored 1 if correct or 0 if incorrect. The distribution of the 100 paired responses is shown below:

		Y:	
		0	1
X:	1	25	50
	0	15	10

Does it make sense to describe the relationship between the responses to these two items as positive?

3.16 A fair die is rolled a large number of times. We count the number of times that the die falls with 1, 2, 3, 4, 5, and 6 dots face up. What would you expect to be approximately true of the shape of this frequency distribution?

3.17 Under what conditions will the shape of a cumulative-proportion graph be a straight line?

3.18 Sketch the shape of a cumulative-proportion graph for a frequency distribution that is skewed to the left.

3.19 Sketch the shape of a cumulative-proportion graph for a frequency distribution that is skewed to the right.

4

Measures of Central Tendency

4.1 Introduction

In the last chapter, we discussed three characteristics of a set of observations. By making a frequency distribution, we can determine how the values of a variable are distributed in the various classes. The frequency distribution enables us to perceive the shape or form of the distribution. The range of the values of the observations, which we defined as the difference between the highest and lowest value in the set, provides us with information concerning the dispersion or variability of the observations. The mode indicates a typical value in the set of observations, that is, the value with the greatest frequency of occurrence.

4.2 The Mean

Measures that describe a typical, representative, or average value for a set of observations are called *measures of central tendency*. The mode is one such measure. Another, and more frequently used, measure of central tendency is the arithmetic mean. The *arithmetic mean* or, more simply, the mean, is defined as the sum of all of the values of the observations, divided by the total number of observations. As a symbol for the mean we shall use a capital letter with a bar over it. For example, we might use $\bar{X}$ to represent the mean. Then, if we have eight observations with values of 7, 6, 5, 4, 4, 3, 2, and 1, we would have

$$\bar{X} = \frac{7+6+5+4+4+3+2+1}{8} = \frac{32}{8} = 4$$

It is convenient to represent a variable of interest by a capital letter. If the variable of interest is the intelligence quotient, as measured by scores on the Stanford-Binet Test, we might designate this variable with the letters IQ. If the variable is a score based on the Rorschach Ink Blot Test, we might use R to indicate this variable. Age, as a variable, might be represented by A, height by H, weight by W, and so forth.

Rather than attempting to use different letters to designate each variable of interest, however, it is more convenient to use a single letter for this purpose. In general, we shall use X (and occasionally Y) to represent any variable of interest and n to represent the number of observations in a given set. Specific individual observations of the variable can be represented by the use of subscripts. Thus, the eight scores could be represented by $X_1, X_2, X_3, \ldots, X_8$ where the subscripts $1, 2, 3, \ldots, 8$ stand for the particular observations. Thus it would be possible for us to represent the mean value of a given variable consisting of $n = 8$ observations as

$$\bar{X} = \frac{X_1 + X_2 + X_3 + \cdots + X_8}{n} \tag{4.1}$$

A symbol that we shall use frequently is Σ, the Greek capital *sigma*. This symbol is an operational as well as a descriptive symbol and means "to sum." Thus ΣX means to "sum the variable X," or simply "summation X," or "sum of the X's." With the use of this symbol, we now define the mean as

$$\bar{X} = \frac{\Sigma X}{n} \tag{4.2}$$

where $\bar{X} =$ the arithmetic mean
  $\Sigma X =$ the sum of the values of the variable
  $n =$ the number of observations in the set.

In Table 4.1, we have arranged the $n = 8$ observations in the form of a frequency distribution. All of the values have a frequency of 1 except the value of $X = 4$, which has a frequency of 2. If we multiply each value by its corresponding frequency to obtain fX and then sum the resulting products, it is obvious that ΣfX will be equal to the sum of the n values of X. For example, for the frequency distribution shown in Table 4.1, we have

$$\Sigma fX = (1)(7) + (1)(6) + (1)(5) + (2)(4) + (1)(3) + (1)(2) + (1)(1)$$
$$= 32$$

which is, of course, equal to ΣX. Consequently, for a frequency distribution, we have

$$\bar{X} = \frac{\Sigma fX}{n} = \frac{\Sigma X}{n} \tag{4.3}$$

Table **4.1** — Frequency Distribution of 8 Scores

(1) X	(2) f	(3) fX
7	1	7
6	1	6
5	1	5
4	2	8
3	1	3
2	1	2
1	1	1
Σ	8	32

We pointed out previously that, in making a frequency distribution, we may group the observations into classes such that the size or width of the class interval is greater than 1. For example, in Table 4.2 we have observations grouped into classes where the size of the class interval is 5. If we have to choose a single value to represent the observations in a given class, the value we select is the midpoint of the class interval, under the assumption that the mean of all observations in a given class is equal to the midpoint of that class. The range of values in the first interval is from 29.5 up to 34.5, and the midpoint of this class interval is 32. Similarly, the midpoints of the various other class intervals are given in column (3). The midpoints constitute a new and more limited set of values of the variable. If we multiply each of these new values of X by the corresponding frequencies given in column (2), we obtain the products of fX shown in column (4). Summing the entries in column (4) we have $\Sigma fX = 6010$. We also have $n = \Sigma f = 100$ observations. Then, substituting in (4.3)

Table **4.2** — Computation of the Arithmetic Mean from a Frequency Distribution

(1) Scores	(2) f	(3) X	(4) fX
85–89	2	87	174
80–84	1	82	82
75–79	4	77	308
70–74	9	72	648
65–69	13	67	871
60–64	26	62	1612
55–59	19	57	1083
50–54	12	52	624
45–49	8	47	376
40–44	3	42	126
35–39	2	37	74
30–34	1	32	32
Σ	100		6010

we obtain

$$\bar{X} = \frac{6010}{100} = 60.1$$

as the mean of the distribution.

4.3 The Median and Other Centiles

The mean is the measure used most frequently to describe the central tendency of a group of observations. Another measure of central tendency that we may have occasion to use, however, is the median. The *median* may be defined as that point on a scale of measurement above which and below which 50 percent of the observations fall.

Table **4.3** — Cumulative Frequency Distribution

(1) Scores	(2) f	(3) cf	(4) cp
85–89	2	100	1.00
80–84	1	98	0.98
75–79	4	97	0.97
70–74	9	93	0.93
65–69	13	84	0.84
60–64	26	71	0.71
55–59	19	45	0.45
50–54	12	26	0.26
45–49	8	14	0.14
40–44	3	6	0.06
35–39	2	3	0.03
30–34	1	1	0.01

To find the median of a set of measurements, we first arrange the observations in order of magnitude or in terms of a frequency distribution, as in Table 4.3. For this distribution the cumulative proportions, given in column (4), show that the median must fall within the class interval 60–64. The reason for this is that 45 percent of the observations fall below this interval and 26 percent within the interval, and we need to find the point below which 50 percent fall. The median must, therefore, fall someplace within the class interval 60–64. To find the point within the interval corresponding to the median, we make use of the following formula:

$$Mdn = l + \left(\frac{0.50n - cf_b}{f_w}\right)i \qquad (4.4)$$

where Mdn = the median

l = the lower limit of the interval in which the median falls

n = the number of observations in the set

cf_b = the sum of the frequencies or the number of observations
below the interval in which the median falls
f_w = the frequency or number of observations within the interval
containing the median
i = the size of the class interval

Substituting in (4.4), we have

$$Mdn = 59.5 + \left[\frac{(0.50)\,(100) - 45}{26}\right]5$$

$$= 60.46$$

as the value of the median.

In calculating the median, we assume that the observations in the
class interval in which the median falls are equally spaced throughout the
interval. The median can be used as a measure of central tendency for any
set of ordered observations.

Figure 4.1 is the cumulative-proportion graph for the data of Table 4.3.
The figure shows how we might also find the value of the median from a
cumulative-proportion graph. The median, as found from the graph, is
approximately 60.5, a value that does not differ too greatly from the value
we obtained by calculation.

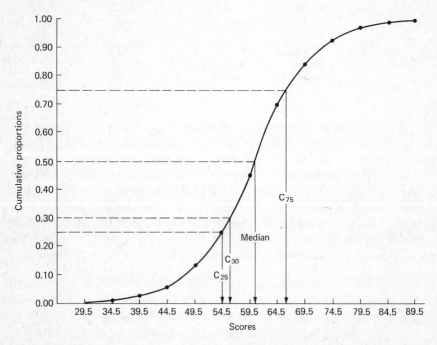

Figure **4.1** — Cumulative-proportion graph for the distribution of scores in Table 4.3.

The median, defined as the point below which and above which 50 percent of the observations fall on the measurement scale, is also called the 50th *centile*, which is symbolized by C_{50}. The 50th centile and the median have the same numerical value. Other centiles could also be found from the cumulative-proportion graph. C_{30}, for example, would be the point on the measurement scale below which 30 percent of the observations fall. Inspection of the graph indicates that C_{30} would fall at approximately 55.5 on the measurement scale.

If we wish to calculate C_{30}, or any other centile, we can do so by redefining the terms in (4.4). For example, to calculate C_{30} we would have

$$C_{30} = l + \left(\frac{0.30n - cf_b}{f_w}\right)i$$

where C_{30} = the 30th centile
l = the lower limit of the interval in which C_{30} falls
n = the number of observations in the set
cf_b = the sum of the frequencies or the number of observations below the interval in which C_{30} falls
f_w = the frequency or number of observations within the interval containing C_{30}
i = the size of the class interval

For the distribution in Table 4.3, we see that C_{30} falls in the interval 55–59. Then, substituting in the above equation, we have

$$C_{30} = 54.5 + \left[\frac{(0.30)(100) - 26}{19}\right]5$$

$$= 55.55$$

In the graph, we show also the values of the 25th and 75th centiles. These two centiles are called the first and third *quartiles* and are sometimes designated by the symbols Q_1 and Q_3, respectively.

4.4 A Proportion as a Mean

In dealing with observations falling into one of two classes, such as the number of Right and Wrong responses to an item on a test, we had occasion to refer to the proportion of the total number of observations falling in a given class. For this nominal scale, we can assign the value of 1 to one of the two classes and the value of 0 to the other. We can think of a set of n such observations as being arranged in the manner of Table 4.4 in which we have assigned the value of $X = 1$ to a Right response and the value of $X = 0$ to a Wrong response. If we use (4.3) to find the mean of this distribution, we get $\Sigma fX = (20)(1) + (30)(0) = 20$ and $\bar{X} = 20/50 = 0.40$. In this case, the mean of the distribution is also the proportion of the total

Table 4.4 — Frequency Distribution of a Variable with Only Two Classes

	(1) f	(2) X	(3) fX
Right	20	1	20
Wrong	30	0	0
Σ	50		20

number of observations in the class assigned the value of $X = 1$. We see that for a variable in which the observations can take only the value of $X = 1$ or $X = 0$, then

$$\bar{X} = \frac{\Sigma X}{n} = \frac{f_1}{n} = p \qquad (4.5)$$

where f_1 = the number of observations assigned the value of 1

p = the proportion of the total number of observations assigned the value of 1

4.5 The Mean of a Set of Ranks

Suppose we have a set of observations and we assign the successive integers from 1 to n to the observations, with 1 being assigned to the largest value, 2 to the next largest, 3 to the next largest, and so on, with the smallest value being assigned n. We refer to this new scale as a ranking or ordinal scale. Table 4.5 gives the values of 7 observations and the corresponding ranks assigned to these values. The sum of the ranks is 28, and the mean is, therefore, $28/7 = 4$.

Table 4.5 — Ranks Assigned to Seven Scores

Scores	Ranks
68	1
62	2
50	3
38	4
25	5
20	6
15	7

Actually, we never have to compute the sum or mean of any set of ranks by adding the various values. Instead, the sum of the ranks from 1 to n is always given by

$$\Sigma X = \frac{n(n+1)}{2} \qquad (4.6)$$

and because the mean is simply the sum divided by n, the mean will always be given by

$$\bar{X} = \frac{n+1}{2} \qquad (4.7)$$

Substituting in these two formulas with n equal to 7, we have

$$\Sigma X = \frac{7(7+1)}{2} = 28$$

and

$$\bar{X} = \frac{7+1}{2} = 4$$

4.6 Some Characteristics of the Various Measures of Central Tendency

4.6.1 THE MODE

The mode has been defined as the value of the observation that occurs most frequently in a given set of observations. If the observations consist of a set of ranks, then the mode does not exist because, in this instance, we have one and only one observation for each of the possible ranks. In the case of a continuous variable in which the recorded or observed values are discrete, it is possible to have more than one observation with the same recorded value. The mode, in this instance, is simply the recorded value that occurs most frequently. If we group observations into broader classes, the mode is taken as the midpoint of the class interval with the greatest frequency. With categorical data, or data classified according to a nominal scale, the mode is the class or category with the greatest frequency.

It is also possible that we may have more than one mode. Figure 4.2 illustrates distributions that would be described as unimodal (one mode), bimodal (two modes), and multimodal (more than two modes). It is only in the case of a unimodal distribution that the mode has meaning as a measure of central tendency.

4.6.2 THE MEDIAN

The median can be used as a measure of central tendency for any scale other than a nominal scale. If the median is to be used as a measure of central tendency, the only essential characteristic of a scale is that the observations be ordered. Thus, the median can be used to describe observations that fall on an ordinal, interval, or ratio scale. A more general definition of a median than one we gave earlier is that given by Wallis and

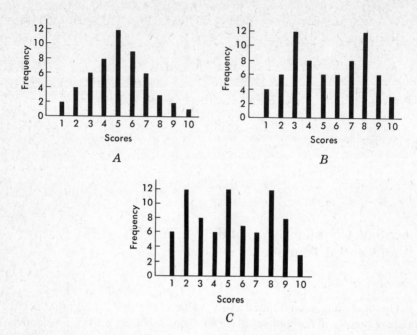

Figure **4.2** — A: A unimodal distribution. B: A bimodal distribution. C: A multimodal distribution.

Roberts.[1] They define the median as any number that does not exceed nor is exceeded by more than half of the observations in a given set. This definition is obviously applicable to ordinal scales as well as interval and ratio scales.

4.6.3 THE MEAN

If we subtract the mean from each of the values in a given set of observations and add the remainders, we would find that this sum is equal to 0. For example, suppose $n = 8$ observations have values of 7, 6, 5, 4, 4, 3, 2, and 1. The sum of these values is 32 and the mean is equal to $32/8 = 4$. Subtracting the mean of 4 from each of the values, we have 3, 2, 1, 0, 0, -1, -2, and -3, and the sum of these deviations is equal to 0.

We shall find it useful, in later discussions, to have a symbol to represent the deviation of a value from the mean. The symbol we shall use for this purpose is x. Thus

$$x = X - \bar{X} \tag{4.8}$$

[1]W. A. Wallis and H. V. Roberts. *Statistics: A new approach.* New York.: The Free Press, 1956, p. 216.

The fact that the sum of the deviations from the mean is equal to 0, we express in the following formula:

$$\Sigma(X - \bar{X}) = \Sigma x = 0 \qquad (4.9)$$

If we alter or change the value of any single observation in a given set, we shall also change the value of the mean of that distribution. The mean is sensitive to each of the numerical values of a variable, because the sum of the deviations from the mean must always equal 0. Thus, if we raise or lower the value of any single observation, we shall accordingly influence the value of the mean. If we increase the value of an observation that falls above the mean, this change will pull the mean toward the higher value. If we decrease the value of an observation that falls below the mean, we shall pull the mean toward the lower value. On the other hand, if we make these same changes for a given set of observations, the median will not be influenced by the changed values. If 50 percent of the observations fall below a value of the variable equal to 60, then 50 percent of the observations will still fall below the value of 60, even though we increase the numerical values of all observations above 60. The median, we may say, is insensitive to extreme values of a variable, whereas the mean is sensitive to these extreme values.

Another property of the mean is that the sum of squared deviations $\Sigma(X - \bar{X})^2$ is at a minimum. That is, if we take some value other than the mean and subtract it from each of the values in a given set and square the resulting deviations, the sum of these squared deviations will always be larger than the sum of squared deviations from the mean. For example, if we have five observations with values of 5, 4, 3, 2, and 1, the mean is 3 and the sum of squared deviations is 10. See if you can find any number other than $\bar{X} = 3$ such that the sum of squared deviations from this number is less than 10 for the given set of five values.

If we know the mean of a set of n observations, we can also find the sum of the values of the variable for this set. This is obviously true from the fact that if $\bar{X} = \Sigma X/n$, then $n\bar{X} = \Sigma X$. This relationship between the number of observations times the mean and the sum of the values enables us, if we have two or more sets of observations of the same variable, to find the mean of the combined sets. All we need to know is the number of observations in each set and the mean of each set.

For example, if we have three means of 4.0, 6.0, and 6.0, with n's of 3, 4, and 5, respectively, the corresponding values of ΣX would be $(3)(4.0) = 12.0$, $(4)(6.0) = 24.0$, and $(5)(6.0) = 30.0$. The sum of the combined set of $3 + 4 + 5 = 12$ observations is, therefore, $12.0 + 24.0 + 30.0 = 66.0$, and the mean of the set of 12 observations is $66.0/12 = 5.5$.

We can represent what we have done above for the general case where we may have any number of means. We use subscripts to indicate the number of observations in a given set and also the mean of a given set.

Thus if we have k sets, we let $n_1, n_2, \ldots, n_k$ represent the number of observations in the successive sets and $\bar{X}_1, \bar{X}_2, \ldots, \bar{X}_k$ represent the corresponding means. We know that the total sum will be equal to $\Sigma X = \Sigma X_1 + \Sigma X_2 + \cdots + \Sigma X_k$ and that the total number of observations will be equal to $n = n_1 + n_2 + \cdots + n_k$. Then the mean of the combined sets will be given by

$$\bar{X} = \frac{\Sigma X}{n} = \frac{n_1 \bar{X}_1 + n_2 \bar{X}_2 + \cdots + n_k \bar{X}_k}{n_1 + n_2 + \cdots + n_k} \qquad (4.10)$$

In general, in all of the later chapters in this text, the mean will be the commonly used measure of central tendency.

REVIEW OF TERMS

bimodal distribution	midpoint of a class interval
centile	mode
first quartile	proportion
mean	third quartile
median	unimodal distribution

REVIEW OF SYMBOLS

X	$\bar{X}$	p
C_{50}	i	n
cp	x	Q_3
Mdn	Q_1	Σ

QUESTIONS AND PROBLEMS

4.1 For the frequency distribution of Problem 3.1, find: (a) the value of the mean and (b) the value of the median.

4.2 We have 160 students who answered an item on a test with the correct response and 40 who answered the item incorrectly. We assign a value of $X = 1$ to each observation that is a correct response and a value of $X = 0$ to each observation that is an incorrect response. Make a frequency distribution for the two values of X and use this distribution to find ΣX and $\bar{X}$. The value of the mean you obtain should be equal to p, the proportion of students in the sample who answered the item correctly.

4.3 We have observations consisting of a set of ranks from 1 to 10. Add the values of the ranks and divide by $n = 10$ to find the mean. The value of the sum that you obtain should be equal to $n(n+1)/2$ and the value of the mean that you obtain should be equal to $(n+1)/2$.

4.4 Suppose we have the following values of a variable X: 1, 2, 3, 4, 4, 5, 6, and 7. (a) Find the mean. (b) Add 1 to each value of X and find the mean; that is, find the mean of $X + 1$. (c) Add 2 to each value of X and find the mean; that is, find the mean of $X + 2$. (d) Are you ready to generalize about what influence adding a constant to each value of a variable will have on the mean?

If so, predict what the value of the mean will be if you add 3 to each value of X; that is, what will the mean of $X+3$ be equal to in the example cited?

4.5 The scores given below were obtained from a class in introductory psychology on an examination. Find the value of (a) the mean and (b) the median.

Scores	f
42–44	2
39–41	3
36–38	4
33–35	4
30–32	6
27–29	10
24–26	15
21–23	11
18–20	10
15–17	5
12–14	3
9–11	4
6–8	2
3–5	1

4.6 Give a one-sentence definition of each of the following concepts: (a) centile, (b) median, (c) mean.

4.7 Suppose we have the following values of a variable X: 1, 2, 3, 4, 4, 5, 6, and 7. You have already found the mean of these eight observations in Problem 4.4. (a) Multiply each value of X by 2 and find the mean; that is, find the mean of $2X$. (b) Multiply each value of X by 3 and find the mean; that is, find the mean of $3X$. (c) Are you ready to generalize about what influence multiplying each value of a variable by a constant will have on the mean? If so, predict what the mean of $4X$ will be in the example cited.

4.8 The mean and number of observations for each of four samples are given below.

	n	$\bar{X}$
Sample 1	10	8.0
Sample 2	5	10.0
Sample 3	10	7.0
Sample 4	20	9.0

What is the value of the mean for the combined set of $n = 45$ observations?

4.9 We have three samples from a binomial population. The mean and number of observations for each sample are given below.

	n	p
Sample 1	20	0.80
Sample 2	10	0.70
Sample 3	40	0.60

What is the mean for the combined set of $n = 70$ observations?

4.10 Suppose we have the following values of a variable X: 1, 2, 3, 4, 4, 5, 6, and 7. On the basis of what you discovered in Problem 4.4, predict what the mean would be if: (a) you subtracted 1 from each value of X, (b) if you subtracted 2 from each value of X, (c) if you subtracted 4 from each value of X.

4.11 Can you prove that if X is a variable and that if you add a constant a to each value of X that the mean of $X + a$ will be equal to $\bar{X} + a$?

4.12 Suppose that we have the following values of a variable X: 1, 2, 3, 4, 4, 5, 6, and 7. On the basis of what you discovered in Problem 4.7, predict what the mean would be if: (a) you multiplied each value of X by 1/2, that is, if you divided each value of X by 2; (b) you multiplied each value of X by 1/4, that is, if you divided each value of X by 4.

4.13 Can you prove that if X is a variable and that if you multiply each value of X by a constant b, then the mean of bX will be equal to $b\bar{X}$?

4.14 For which of the four types of scales do each of the following measures of central tendency make sense: (a) the mode, (b) the mean, and (c) the median? Explain why.

4.15 Can you prove that $\Sigma(X - \bar{X}) = 0$?

4.16 Under what conditions does it make sense to describe the mean for one sample as being larger than the mean for another sample?

4.17 The mean and number of observations for each of four samples are given below. What is the value of the mean for the combined set of $n = 45$ observations?

	n	$\bar{X}$
Sample 1	5	9.0
Sample 2	10	8.0
Sample 3	20	6.0
Sample 4	10	7.0

4.18 Calculate the 50th centile of the following distribution:

X	f
9	1
8	2
7	3
6	4
5	5
4	4
3	3
2	2
1	1

4.19 The value of the mean of the distribution given in Problem 4.18 can be determined by inspection. Explain why. What is the value of the mean?

4.20 For a sample of n observations, we have $\bar{X} = 10.0$. We change each value of X in the various ways shown below to obtain a new variable Y.

(a) $Y = 2X$
(b) $Y = X - 2$
(c) $Y = X/2$
(d) $Y = X + 2$
(e) $Y = (X - 10)/2$

For each of the above, what will be the value of $\bar{Y}$?

4.21 Let a be a constant. Then, if $Y = X - a$, what value of a will make $\Sigma Y^2 = \Sigma (X - a)^2$ a minimum?

4.22 Explain why the mean of the three numbers 7, 8, and 9, is exactly equal to the mean of each of the following sets of numbers:

(a) 6, 8, 10
(b) 5, 8, 11
(c) 12, 8, 4
(d) 17, 8, −1

5

Measures of Variability

5.1 Introduction

The *range* is one measure of dispersion or variability within a given set of observations. It is not, however, the measure that we shall have occasion to use most frequently. Rather, we shall prefer a measure that is called the *standard deviation*, or its square, which is called the *variance*. The standard deviation and variance are based on the sum of *squared deviations* from the mean of a set of observations.

5.2 The Variance and Standard Deviation

We illustrate the calculation of the variance and the standard deviation with the following set of $n = 8$ values of a variable: 7, 6, 5, 4, 4, 3, 2, and 1, given in Table 5.1. The sum of these values is 32 and the arithmetic mean is $32/8 = 4$. Subtracting the mean from each value, we have 3, 2, 1, 0, 0, −1, −2, and −3, as shown in column (2) of the table. Squaring each of these deviations, we obtain 9, 4, 1, 0, 0, 1, 4, and 9, as shown in column (3) of the table. The sum of the squared deviations is 28 and, if we divide the sum by $n − 1$, that is, by one less than the number of observations in the set, we get $28/7 = 4$.

The statistic we have just calculated is called the variance and is symbolized by s^2. In terms of the symbols we have defined previously, we have

$$s^2 = \frac{\Sigma (X - \bar{X})^2}{n - 1} \qquad (5.1)$$

Table **5.1** — Computation of the Sum of Squared Deviations

	(1) X	(2) $X - \bar{X}$	(3) $(X - \bar{X})^2$	(4) X^2
	7	3	9	49
	6	2	4	36
	5	1	1	25
	4	0	0	16
	4	0	0	16
	3	−1	1	9
	2	−2	4	4
	1	−3	9	1
Σ	32	0	28	156

If we take the square root of the variance, we obtain the standard deviation or

$$s = \sqrt{\frac{\Sigma(X - \bar{X})^2}{n-1}} \tag{5.2}$$

which, for the set of $n = 8$ observations, gives

$$s = \sqrt{\frac{28}{8-1}} = 2$$

5.2.1 THE SUM OF SQUARED DEVIATIONS

Because the variance and standard deviation are used frequently as measures of variability, we illustrate various methods for calculating them, as we did in the case of the mean. The essential quantity we need to calculate is $\Sigma(X - \bar{X})^2$, which we call a *sum of squared deviations*. We now note that

$$\Sigma(X - \bar{X})^2 = \Sigma X^2 - \frac{(\Sigma X)^2}{n} \tag{5.3}$$

The right side of (5.3) is obtained by expanding the left side and summing and substituting an identity, $\Sigma X/n$, for $\bar{X}$ in the resulting expression.

Now look at Table 5.1. Note that we have entered in column (1) values of X and in column (4) values of X^2. The sums of these two columns are $\Sigma X = 32$ and $\Sigma X^2 = 156$, respectively. Substituting in (5.3) with these two values, we have

$$\Sigma(X - \bar{X})^2 = 156 - \frac{(32)^2}{8}$$

$$= 28$$

as before.

Table **5.2** — Computation of the Sum of Squared Deviations from a Frequency Distribution with Class Interval Equal to 1

(1) X	(2) f	(3) fX	(4) fX^2
7	1	7	49
6	1	6	36
5	1	5	25
4	2	8	32
3	1	3	9
2	1	2	4
1	1	1	1
Σ	8	32	156

If the same observations are arranged in the form of a frequency distribution, as in Table 5.2, a similar formula would apply. In this table we give in column (3) the values of fX and in column (4) the values of fX^2. Summing both columns, we obtain $\Sigma fX = 32$ and $\Sigma fX^2 = 156$. Then

$$\Sigma(X - \bar{X})^2 = \Sigma fX^2 - \frac{(\Sigma fX)^2}{n} \tag{5.4}$$

or, for our example,

$$\Sigma(X - \bar{X})^2 = 156 - \frac{(32)^2}{8} = 28$$

as before.

Table **5.3** — Computation of the Sum of Squared Deviations from a Frequency Distribution with Class Interval Greater than 1

(1) Scores	(2) f	(3) X	(4) X^2	(5) fX	(6) fX^2
85–89	2	87	7569	174	15,138
80–84	1	82	6724	82	6,724
75–79	4	77	5929	308	23,716
70–74	9	72	5184	648	46,656
65–69	13	67	4489	871	58,357
60–64	26	62	3844	1612	99,944
55–59	19	57	3249	1083	61,731
50–54	12	52	2704	624	32,448
45–49	8	47	2209	376	17,672
40–44	3	42	1764	126	5,292
35–39	2	37	1369	74	2,738
30–34	1	32	1024	32	1,024
Σ	100			6010	371,440

If we have a frequency distribution in which the class interval is greater than 1, then we can use the midpoints of the class intervals, as we did previously in calculating the mean, to find the sum of squares. The necessary calculations are illustrated in Table 5.3. Table 5.3 uses the same data as Table 4.2 of the previous chapter, but we have added a column headed X^2 and another headed fX^2. The values of X^2 are simply the squares of the midpoints of the class intervals and were easily obtained from Table II in the Appendix. Each value of X^2 is then multiplied by its corresponding frequency f and the products of fX^2 are entered in column (6) of Table 5.3. Summing column (5) we have $\Sigma fX = 6010$ and summing column (6) we have $\Sigma fX^2 = 371{,}440$. Then, for this distribution we have, by substitution in (5.4),

$$\Sigma (X - \bar{X})^2 = 371{,}440 - \frac{(6010)^2}{100} = 10{,}239$$

For the variance and standard deviation of the distribution, we have

$$s^2 = \frac{10{,}239}{100 - 1} = 103.42$$

and

$$s = \sqrt{\frac{10{,}239}{100 - 1}} = 10.2$$

We have said that both the variance and the standard deviation are useful measures of the variability of a set of observations. Later we shall see why they are useful. For the time being, it is important only that you be able to calculate the sum of squared deviations, which is necessary in order to obtain the value of the variance and of the standard deviation.

5.3 Statistics and Parameters

When we have a set of n observations, in many instances we wish to use the values of these observations to estimate certain characteristics of a larger number of observations that we have not made. Our interest, in other words, is not directly in a particular set of observations at hand, but rather in using them to estimate something about a potentially larger set we could make but do not have at hand. We shall refer to the observations we have made as a *sample*. The larger supply, which we do not have, we shall call a *population*. From the characteristics of a sample of observations, we are often interested in estimating similar characteristics for the population.

When we calculate some measure descriptive of a sample, we call this value a *statistic*. A statistic is any characteristic of a set of observations calculated from a sample. The measure corresponding to a statistic when obtained from a population, we refer to as a *parameter*. In most cases of interest, the values of parameters are unknown and must be estimated by the statistics derived from a sample. Thus the sample mean is an estimate of the population mean, and the variance, as given by (5.1), is a characteristic of a sample that is also an estimate of the population variance.

When we refer to the mean of a sample, we use the symbol $\bar{X}$. To represent the population mean, we use the symbol μ. When we refer to the standard deviation and variance of a sample, we use the symbols s and s^2, respectively. The corresponding values in the population will be indicated by σ and σ^2, respectively. For the number of observations in a sample, we use n, and for the corresponding number in a population, we use N.

In general, knowledge of the value of a population mean tells us nothing about the variability of the individual observations in the population. Any number of populations may have identical means and still differ considerably in variability. Similarly, any number of populations may have equal variances and still differ considerably with respect to their means.

5.4 The Mean and Standard Deviation of a Binomial Population

We have said that, in general, if we know the value of a population mean, this fact does not convey any information about the variability of the individual observations in the population. An exception to this statement is a population in which a variable can take only the values of $X = 1$ or $X = 0$. This population is called a *binomial population*. If we know the mean of a binomial population, the standard deviation and variance are also known exactly. Knowing the value of the population mean, in other words, enables us to determine precisely the population standard deviation and variance.

Assume, for example, that a population contains N observations of which F_1 observations have the value of $X = 1$ and F_0 observations have the value of $X = 0$. We use the symbol F rather than f, in this instance, to indicate that we are dealing with population frequencies rather than sample frequencies. Then the total number of observations in the population will be equal to $N = F_1 + F_0$. We use N rather than n to represent the number of observations in the population. Now, obviously all of the F_0 observations with $X = 0$ will contribute nothing to ΣX. Thus, ΣX will simply be the number of observations with $X = 1$ or, in other words, $\Sigma X = F_1$. Then the population mean will be given by

$$\mu = \frac{\Sigma X}{N} = \frac{F_1}{N} \tag{5.5}$$

where we have used μ rather than $\bar{X}$ because we are dealing with a population mean rather than a sample mean.

We let $P = F_1/N$, or the proportion of observations in the population with $X = 1$, and we note that for a binomial population

$$\mu = \frac{\Sigma X}{N} = P$$

As a simple illustration, assume that we have $F_1 = 3$ observations with $X = 1$ and $F_0 = 2$ observations with $X = 0$. Then

$$\mu = \frac{3}{5} = 0.6$$

Note also that for this binomial population, we have

$$\Sigma X^2 = (1)^2 + (1)^2 + (1)^2 + (0)^2 + (0)^2 = F_1$$

or simply the number of observations for which $X = 1$. In general, for any binomial population, we have

$$\Sigma X^2 = F_1$$

The sum of squared deviations from the population mean μ will be given by

$$\Sigma (X - \mu)^2 = \Sigma X^2 - \frac{(\Sigma X)^2}{N} \qquad (5.6)$$

but we have already shown that $\Sigma X^2 = F_1$ and that $\Sigma X = F_1$, and substituting in (5.6) with these identities, we have

$$\Sigma (X - \mu)^2 = F_1 - \frac{(F_1)^2}{N}$$

We divide the sum of squares, defined above, by N, the number of observations in the population, and this will give us the population variance. We divide by N rather than by $n - 1$ because we are dealing with a population rather than a sample, and we use the symbol σ^2 instead of s^2 for the same reason.[1] Thus

$$\sigma^2 = \frac{F_1 - (F_1)^2/N}{N}$$

$$= \frac{F_1}{N} - \frac{(F_1)^2}{N^2}$$

$$= P - P^2$$

---

[1] When deviations are taken from the mean of a random sample rather than from a population mean, it can be shown that division of the sum of squared deviations by $n - 1$ gives an unbiased estimate of the population variance, whereas division by n does not. Division of the sum of squared deviations by n or by $n - 1$ would, of course, serve to describe the variance of the sample. However, since the unbiased estimate of the population variance plays an important role in statistical inference, and since it is an equally good description of the sample variance, we have consistently divided the sample sum of squared deviations by $n - 1$.

When the complete population of N observations is available, as in the case described above, there is no question of obtaining an unbiased estimate of the population variance. The parameter can be obtained from the data.

or

$$\sigma^2 = P(1-P) \qquad (5.7)$$

It will be convenient to let $Q = 1 - P$. Then for any binomial population we have

$$\sigma^2 = PQ \qquad (5.8)$$

and

$$\sigma = \sqrt{PQ} \qquad (5.9)$$

For our example in which $F_1 = 3$ and $F_0 = 2$, we have $P = 3/5 = 0.6$ and $Q = 1 - P = 0.4$. Then, for this population, we have

$$\sigma^2 = (0.6)(0.4) = 0.24$$

and

$$\sigma = \sqrt{0.24} = 0.4899$$

We thus see that for this distribution, and all others of this kind, if the population mean is known, then the population variance and standard deviation are completely determined by the value of the mean.[2]

For purposes of illustration, we show in Table 5.4 the $N = 5$ values of X for which we have $F_1 = 3$ observations with $X = 1$ and $F_0 = 2$ observations with $X = 0$. It is obvious that $\mu = P = 3/5 = 0.6$. The deviations, $X - \mu$, are shown in column (2) of the table and the squared deviations, $(X - \mu)^2$, are shown in column (3). Note that $\sigma^2 = \Sigma(X - \mu)^2/N$ is equal to $1.20/5 = 0.24$ and that this is exactly equal to $\sigma^2 = P(1 - P)$ or $(0.6)(1 - 0.6) = 0.24$.

Table 5.4 — A Binomial Population with $P = 0.6$ and with $N = 5$

	(1) X	(2) $X - \mu$	(3) $(X - \mu)^2$
	1	0.4	0.16
	1	0.4	0.16
	1	0.4	0.16
	0	-0.6	0.36
	0	-0.6	0.36
Σ	3	0.0	1.20

5.5 The Mean and Standard Deviation of a Population of Ranks

Suppose that we have a population that consists of a set of ranks from 1 to N. In Table 5.5, we consider a population with ranks from 1 to 7. The

---

[2]It is important to note that the variance given by (5.8) and (5.9) is independent of N, the number of observations in the population. All binomial populations with the same value of P will have the same variance, regardless of the number of observations in the population.

Table **5.5** — Computation of the Sum of Squared Deviations for a Population of $N = 7$ Ranks

	(1) X	(2) $X - \mu$	(3) $(X - \mu)^2$
	7	3	9
	6	2	4
	5	1	1
	4	0	0
	3	−1	1
	2	−2	4
	1	−3	9
Σ	28	0	28

mean of a population of N ranks will be given by

$$\mu = \frac{\Sigma X}{N} = \frac{N+1}{2} \tag{5.10}$$

and for the ranks given in Table 5.5, we have

$$\mu = \frac{28}{7} = \frac{7+1}{2} = 4$$

In column (2) we give the deviations, $X - \mu$, and in column (3) the squared deviations, $(X - \mu)^2$. The population variance will be given by

$$\sigma^2 = \frac{\Sigma(X - \mu)^2}{N} \tag{5.11}$$

and for the data of Table 5.5, we have $\Sigma(X - \mu)^2 = 28$. Substituting in (5.11), we have

$$\sigma^2 = \frac{28}{7} = 4$$

In general, for a set of N ranks, it is not necessary to calculate $\Sigma(X - \mu)^2$ directly because it can be shown that for any set of N ranks

$$\Sigma(X - \mu)^2 = \frac{N^3 - N}{12} \tag{5.12}$$

For the $N = 7$ ranks we have

$$\Sigma(X - \mu)^2 = \frac{7^3 - 7}{12} = 28$$

and this is exactly equal to the value shown in Table 5.5. We also have for any population of N ranks

$$\sigma^2 = \frac{\Sigma(X - \mu)^2}{N} = \frac{(N^3 - N)/12}{N} = \frac{N^2 - 1}{12} \tag{5.13}$$

and

$$\sigma = \sqrt{\frac{N^2 - 1}{12}} \qquad (5.14)$$

or, for our example,

$$\sigma = \sqrt{\frac{7^2 - 1}{12}} = 2$$

We note that for any population of N ranks, the mean and variance are determined completely by N, the number of ranks.

5.6 Other Measures of Variability

The *average* or *mean deviation* is sometimes used as a measure of variability. The mean deviation is defined as the sum of the absolute differences between the values of the observations and the arithmetic mean, divided by the number of observations. We cannot simply sum the deviations from the mean, because this sum is always equal to zero. When we say we sum the absolute values of the differences, we mean that we ignore the signs of the deviations. We indicate the absolute value of a deviation by $|X - \bar{X}|$. Then the average deviation is defined as

$$AD = \frac{\Sigma |X - \bar{X}|}{n} \qquad (5.15)$$

If we have $n = 8$ values consisting of 7, 6, 5, 4, 4, 3, 2, and 1, the mean is equal to 4. The deviations of the values of X from the mean are 3, 2, 1, 0, 0, -1, -2, and -3. Ignoring the signs of the deviations and summing the absolute values we have

$$AD = \frac{3 + 2 + 1 + 0 + 0 + 1 + 2 + 3}{8} = 1.5$$

as the value of the average deviation.

Another measure of variability is the middle 80 percent range, or the range of scores between the 90th and 10th centiles. Another is the interquartile range, or the range of scores between the 75th and 25th centiles. Additional measures of variability based on differences between scores corresponding to various centiles could also be used to describe the variability of a given set of observations.

In general, we shall have little use for measures of variability other than the variance and the standard deviation. One reason for preferring these two measures of variability is that they have algebraic properties such that if we have several sets of observations, and if we know the means and standard deviations of each of the sets, we can obtain the variance and standard deviations of the combined set of observations. Another reason is that these measures have played an important role in statistical theory with reference to a particular form of distribution known as the *normal distribution*. We shall see the usefulness of the normal distribution and the standard deviation of this distribution in a variety of problems in later discussions.

REVIEW OF TERMS

average deviation	ranks
binomial population	sample
interquartile range	standard deviation
parameter	statistic
population	variance

REVIEW OF SYMBOLS

The following table summarizes the symbols used when we are dealing with samples and with populations.

	Symbols	
	Sample	*Population*
Mean	$\bar{X}$	μ
Variance	s^2	σ^2
Standard deviation	s	σ
Number of observations	n	N
Frequency	f	F
Proportion	p	P

QUESTIONS AND PROBLEMS

5.1   For the frequency distribution of Problem 3.1 find: (a) the value of the variance and (b) the value of the standard deviation.

5.2   Suppose we have a population that consists of ranks from 1 to 9. (a) Subtract the population mean from each rank, square the resulting deviations, and sum. You should find that this sum of squared deviations is equal to $(N^3 - N)/12$, where N is equal to the number of ranks. (b) If you divide the sum of squared deviations by N to obtain the population variance σ^2, you should find that σ^2 is equal to $(N^2 - 1)/12$, where N is equal to the number of ranks.

5.3   Assume that 160 students answer an item on a test correctly and 40 do not. Let us assume that the $N = 200$ students constitute a population. Assign the value of $X = 1$ to those observations consisting of a correct response and $X = 0$ to those that are incorrect. (a) Make a frequency distribution for these two values of X and find the value of the population mean. You should find that the mean of this distribution is equal to P, the proportion of observations in the population with $X = 1$. (b) Subtract the population mean from each value of X and find $\Sigma F(X-\mu)^2/N$. This will be the population variance σ^2 and you should find that σ^2 is equal to PQ, where P is the proportion of observations in the population with $X=1$ and $Q=1-P$.

5.4   What is the essential difference between a statistic and a parameter?

5.5   What is the difference between a sample and a population?

5.6   Suppose we have the following values of a variable X: 1, 2, 3, 4, 4, 5, 6, and 7. (a) Find s^2, the variance of the set of $n=8$ observations. (b) Add 2 to each value of X and find the variance; that is, find the variance of $X+2$. (c) Add 3 to each value of X and find the variance; that is, find the variance of $X+3$. (d) Are you ready to generalize about what influence an added constant will have upon the variance? If so, predict the variance that you would obtain, if you added 4 to each value of X in the example cited.

5.7   On the basis of what you discovered in Problem 5.6, predict what the variance would be equal to if you subtracted 2 from each value of X in Problem 5.6.

5.8   Suppose we have a variable X with the following values: 1, 2, 3, 4, 4, 5, 6, and 7. (a) Multiply each value of X by 2 and find the variance; that is, find the variance of $2X$. (b) Multiply each value of X by 3 and find the variance; that is, find the variance of $3X$. (c) Are you ready to generalize about what influence multiplying each value of a variable by a constant will have upon the variance? If so, predict what the variance would be equal to if you multiplied each value of X by 4.

5.9   On the basis of what you discovered in Problem 5.8, predict what the variance would be if you: (a) multiplied each value of X by 1/2, that is, divided each value of X by 2, (b) if you multiplied each value of X by 1/4, that is, if you divided each value of X by 4.

5.10 Can you prove that if you multiply each value of a variable by a constant b, the variance of bX will be equal to b^2s^2?

5.11 Can you prove that if you divide each value of X by a constant b, the variance of X/b will be equal to s^2/b^2?

5.12 If we have a population of $N=25$ ranks, what are the values of μ and σ?

5.13 We have one sample with the following values of X: 1, 2, 3, 4, and 5. For another sample the values of X are: 100, 101, 102, 103, and 104. Are the variances for these two samples equal to the same value? Explain why.

5.14 If we have a sample of n observations from a population with known mean μ, will $\Sigma(X-\bar{X})^2$, in general, be greater than or less than $\Sigma(X-\mu)^2$? Explain why.

5.15 We have the following $n=8$ values of X: 8, 9, 10, 11, 11, 12, 13, and 14. Find the value of $\bar{X}$ and the value of s.

5.16 Which, if any, of the statements listed below would you regard as meaningful?

(a) The variance of the heights of males is greater than the variance of their weights.
(b) Females have greater variability in their IQ's than do males.
(c) On a standardized test of reading comprehension, the variance of the scores of 12-year-old girls was found to be larger than the variance of the scores of 12-year-old boys.
(d) Scores of males on a test of reading comprehension have a larger variance than do their IQ's.

5.17 If we have a population of $N = 73$ ranks, what are the values of μ and σ?

5.18 Calculate $\bar{X}$ and s for the following values of X: 8, 8, 6, 3, and 0.

5.19 See if you can prove that for a binomial population

$$\sigma^2 = \frac{\Sigma(X-\mu)^2}{N} = PQ$$

where $P = F_1/N$ and $Q = 1 - P$.

5.20 See if you can prove that $\Sigma(X-\bar{X})^2 = \Sigma X^2 - (\Sigma X)^2/n$.

5.21 See if you can prove that for a binomial population $\mu = P$.

5.22 For one sample we have the following values of X: 2, 4, 6, and 8. For another sample the values of X are: 60, 62, 64, and 66. Are the variances of these two samples equal to the same value? Explain why.

5.23 We have one binomial population with $F_1 = 6$ values of $X = 1$ and $F_0 = 12$ values of $X = 0$ and another binomial population with $F_1 = 33$ values of $X = 1$ and $F_0 = 66$ values of $X = 0$. Are the variances of these two populations equal to the same value? Explain why.

5.24 We can have any number of different binomial populations with different population means $\mu = P$. For which binomial population will the variance, σ^2, be the largest possible value?

5.25 For a sample of n observations we have $\bar{X} = 10$ and $s = 2.0$. We change each value of X in the various ways shown below to obtain a new variable Y.

(a) $Y = 2X$
(b) $Y = X - 2$
(c) $Y = X/2$
(d) $Y = X + 2$
(e) $Y = (X - 10)/2$

For each of the Y variables listed above, give the value of the variance and the standard deviation.

5.26 Is it true that $\Sigma(X-\bar{X})^2 = \Sigma X^2 - n\bar{X}^2$?

6

The Correlation Coefficient

6.1 Introduction

The statistical techniques discussed so far are useful for describing distributions of single variables or univariate distributions. We are now ready to consider statistical techniques that will permit us to study the distribution of the paired values of two variables. Such distributions, as we learned in Chapter 3, are called bivariate distributions. We let one of the variables be represented by X and the other by Y. Each observation consists of a pair of (X,Y) values. Our interest is in the possible *relationship* or *association* that may exist between the values of the two variables.

When large values of one variable are associated with large values of the second variable, and small values of the first with small values of the second, we say that the two variables are *positively* related. When large values of one variable are associated with small values of the second, and small values of the first with large values of the second, we say that the two variables are *negatively* related. Obviously, this description of association in terms of positive and negative is meaningful only if the values of both variables can be ordered. If the values of one or both of the two variables cannot be ordered, they may still be associated. We would not, however, use the terms "positive" and "negative" in describing the nature of the association.

We must take care, in studying the association between two variables, that we do not confuse the concepts of "association" and "causation." When two things are associated, it does not necessarily follow that one is the cause of the other. We might find that there is a relationship between scores on a test of aggressiveness and yearly income for a sample of several hundred men, but we cannot legitimately conclude, on the basis of this finding alone, that one is the cause of the other. We do not know whether earning more money "causes" one to be more aggressive, or whether being more aggressive "causes" one to earn more money. We know only what we have found, namely, that in the sample the two variables are related.

The assumption that changes in one variable are the cause of changes in another variable may or may not be valid, but the validity of this assumption must be determined by considerations other than the mere fact that the two variables are related. The changes in each variable, for example, might possibly be the common result of a third variable that we have not taken into account. Statistical techniques can provide us with an answer to the question: Are the two variables associated? If we find that they are, we still have not answered the question: Why are these two variables related?

6.2 The Covariance

In Table 6.1, we show the values of X and Y in columns (2) and (3) for $n = 8$ observations. Columns (4) and (5) give the corresponding values of $x = X - \bar{X}$ and $y = Y - \bar{Y}$, and columns (6) and (7) the values of $x^2 = (X - \bar{X})^2$ and $y^2 = (Y - \bar{Y})^2$. Column (8) gives the product of the paired (x,y) values. The sum of column (8) is called the *deviation product sum*. The deviation product sum, when divided by $n - 1$, is called the *covariance*. If we let c_{XY} designate the covariance, then

$$c_{XY} = \frac{\Sigma(X - \bar{X})(Y - \bar{Y})}{n - 1} \tag{6.1}$$

or

$$c_{XY} = \frac{\Sigma xy}{n - 1} \tag{6.2}$$

Table **6.1** — Calculation of the Correlation Coefficient: Deviation Scores and Standard Scores

(1) Observation	(2) X	(3) Y	(4) x	(5) y	(6) x^2	(7) y^2	(8) xy	(9) z_X	(10) z_Y	(11) $z_X z_Y$
1	7	19	3	6	9	36	18	1.5	1.5	2.25
2	6	17	2	4	4	16	8	1.0	1.0	1.00
3	5	15	1	2	1	4	2	0.5	0.5	0.25
4	4	13	0	0	0	0	0	0.0	0.0	0.00
5	4	13	0	0	0	0	0	0.0	0.0	0.00
6	3	11	−1	−2	1	4	2	−0.5	−0.5	0.25
7	2	9	−2	−4	4	16	8	−1.0	−1.0	1.00
8	1	7	−3	−6	9	36	18	−1.5	−1.5	2.25
Σ	32	104	0	0	28	112	56	0.0	0.0	7.00

In our example we have

$$c_{XY} = \frac{56}{8-1} = 8$$

6.3 The Correlation Coefficient

As a measure of the degree of relationship between two variables, we consider the *correlation coefficient*. The correlation coefficient may be defined as the covariance of X and Y divided by the square root of the product of the variance of X and the variance of Y. The symbol commonly used to represent the correlation coefficient is r. In terms of the definition we have given, we have

$$r = \frac{c_{XY}}{\sqrt{s_X^2 s_Y^2}} \tag{6.3}$$

or

$$r = \frac{\dfrac{\Sigma (X-\bar{X})(Y-\bar{Y})}{n-1}}{\sqrt{\left[\dfrac{\Sigma (X-\bar{X})^2}{n-1}\right]\left[\dfrac{\Sigma (Y-\bar{Y})^2}{n-1}\right]}} \tag{6.4}$$

or

$$r = \frac{\dfrac{\Sigma xy}{n-1}}{\sqrt{\left(\dfrac{\Sigma x^2}{n-1}\right)\left(\dfrac{\Sigma y^2}{n-1}\right)}} \tag{6.5}$$

Substituting in (6.5) with the appropriate values from Table 6.1, we have

$$r = \frac{\dfrac{56}{8-1}}{\sqrt{\left(\dfrac{28}{8-1}\right)\left(\dfrac{112}{8-1}\right)}} = 1.00$$

as the value of the correlation coefficient for the data of the table.

We note that (6.3) can also be written as

$$r = \frac{c_{XY}}{s_X s_Y} \tag{6.6}$$

and multiplying both sides of (6.6) by $s_X s_Y$ we obtain an important identity for the covariance. Thus

$$c_{XY} = r s_X s_Y \tag{6.7}$$

6.4 Standard-Score Formula for r

From (6.3) we can derive a variety of other formulas for the correlation coefficient. One of these, known as the standard-score formula, is particularly useful for gaining some understanding of the nature of the correlation coefficient but is not very useful in actually calculating the coefficient. The standard-score formula requires that we first translate the original X and Y values into standard scores. We define a *standard score* as

$$z = \frac{X - \bar{X}}{s} \tag{6.8}$$

where z = a standard score
    $\bar{X}$ = the arithmetic mean
    s = the standard deviation

It is obvious that a standard score, as defined by (6.8), is simply a linear transformation of a given measurement X to a new scale. On this new scale, the standard-score scale, the mean of the z values will be equal to 0, and the standard deviation will be equal to 1. We can translate any distribution of scores to a standard-score scale by expressing the original scores as deviations from the sample mean and then dividing the deviations by the sample standard deviation.

For the data of Table 6.1, we have $\bar{X} = 32/8 = 4$, and $s_X = \sqrt{28/(8-1)}$ = 2. For the Y variable, we have $\bar{Y} = 104/8 = 13$ and $s_Y = \sqrt{112/(8-1)} = $ 4. Using these two sample statistics, we obtain the values of z_X given in column (9) of Table 6.1 and the z_Y values given in column (10). The products of the paired (z_X, z_Y) values are given in column (11). Then the correlation coefficient may also be defined as

$$r = \frac{\Sigma z_X z_Y}{n - 1} \tag{6.9}$$

where r = the correlation coefficient
    $z_X = (X - \bar{X})/s_X$
    $z_Y = (Y - \bar{Y})/s_Y$
    n = the number of observations

Substituting in (6.9) with $\Sigma z_X z_Y = 7$, from Table 6.1, we have

$$r = \frac{7}{8 - 1} = 1.00$$

as before.

When an observation (X, Y) is above the mean on both the X and Y variables, then both z_X and z_Y will be positive in sign and the product of the paired values will also be positive in sign. This will be true of all such observations. If an observation (X, Y) is below the mean on both the X and the Y variables, then both z_X and z_Y will be negative in sign, but the product of the paired values will be positive in sign. The sum of the products $z_X z_Y$ will be at its *maximum positive* value when all pairs of (z_X, z_Y) values are numerically equal and have the same sign. Under this condition, the correlation coefficient will take its maximum positive value and will be equal to 1.00. If there is some tendency for the paired (z_X, z_Y) values to be of the same sign, but not necessarily numerically equal, the value of r will be positive but less than 1.00. We describe all positive values of r as indicating a positive relationship between the two variables.

When a negative relationship exists between two variables, this means that an observation that is above the mean on one variable tends to be below the mean on the other variable. Suppose, for example, that the order of the Y values for the $n = 8$ observations of Table 6.1 were reversed, so that for the paired values we now had $(7,7), (6,9), (5,11), \ldots,$ $(1,19)$, as shown in Table 6.2. We now see that the paired (z_X, z_Y) values are numerically equal, but opposite in sign. Under this condition, the correlation coefficient will take its *maximum negative* value and will be equal to -1.00. All negative values of r from -1.00 to 0.00 are taken as indicating that the two variables are negatively related.

Table 6.2 — An Example of Negative Correlation

(1) Observation	(2) X	(3) Y	(4) x	(5) y	(6) z_X	(7) z_Y	(8) $z_X z_Y$
1	7	7	3	−6	1.5	−1.5	−2.25
2	6	9	2	−4	1.0	−1.0	−1.00
3	5	11	1	−2	0.5	−0.5	−0.25
4	4	13	0	0	0.0	0.0	0.00
5	4	13	0	0	0.0	0.0	0.00
6	3	15	−1	2	−0.5	0.5	−0.25
7	2	17	−2	4	−1.0	1.0	−1.00
8	1	19	−3	6	−1.5	1.5	−2.25
Σ	32	104	0	0	0.0	0.0	−7.00

If there is no consistent tendency for the z_X values to be associated with the z_Y values of the same or opposite sign then, in the limiting case, $\Sigma z_X z_Y$ will be equal to zero and so also will the correlation coefficient. Under this circumstance, we would say that the two variables are not related. The scale of measurement for the correlation coefficient is thus from -1.00 to 1.00, as shown in Figure 6.1.

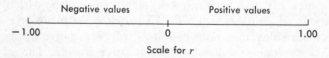

Figure 6.1 — Scale of measurement for the correlation coefficient r.

6.5 Raw Score Formulas for *r*

As we have pointed out, a variety of formulas for calculating the correlation coefficient can be derived from (6.3). We illustrate only a few of the many possible formulas that might be derived. The ones given here involve calculating the correlation coefficient directly in terms of the original measurements without the necessity of first expressing them as deviations, that is, in terms of $x = (X - \bar{X})$ and $y = (Y - \bar{Y})$. In Table 6.3 we repeat the X and Y values of Table 6.1 and also give the values of X^2 and Y^2 and the products XY. We already know that

$$\Sigma(X - \bar{X})^2 = \Sigma X^2 - \frac{(\Sigma X)^2}{n}$$

and that

$$\Sigma(Y - \bar{Y})^2 = \Sigma Y^2 - \frac{(\Sigma Y)^2}{n}$$

and it can also be shown that

$$\Sigma(X - \bar{X})(Y - \bar{Y}) = \Sigma XY - \frac{(\Sigma X)(\Sigma Y)}{n} \tag{6.10}$$

Table 6.3 — Calculation of the Correlation Coefficient Using Original Values of X and Y

(1) Observation	(2) X	(3) Y	(4) X^2	(5) Y^2	(6) XY
1	7	19	49	361	133
2	6	17	36	289	102
3	5	15	25	225	75
4	4	13	16	169	52
5	4	13	16	169	52
6	3	11	9	121	33
7	2	9	4	81	18
8	1	7	1	49	7
Σ	32	104	156	1464	472

Multiplying both the numerator and denominator of (6.4) by $n - 1$, we have

$$r = \frac{\Sigma(X - \bar{X})(Y - \bar{Y})}{\sqrt{[\Sigma(X - \bar{X})^2][\Sigma(Y - \bar{Y})^2]}} \tag{6.11}$$

or

$$r = \frac{\Sigma XY - \dfrac{(\Sigma X)(\Sigma Y)}{n}}{\sqrt{\left[\Sigma X^2 - \dfrac{(\Sigma X)^2}{n}\right]\left[\Sigma Y^2 - \dfrac{(\Sigma Y)^2}{n}\right]}} \tag{6.12}$$

and, substituting in (6.12) with the appropriate values from Table 6.3, we have

$$r = \frac{472 - \dfrac{(32)(104)}{8}}{\sqrt{\left[156 - \dfrac{(32)^2}{8}\right]\left[1464 - \dfrac{(104)^2}{8}\right]}} = 1.00$$

as before.

A variation of (6.12) is obtained by multiplying both the numerator and denominator by n. We then have

$$r = \frac{n\Sigma XY - (\Sigma X)(\Sigma Y)}{\sqrt{[n\Sigma X^2 - (\Sigma X)^2][n\Sigma Y^2 - (\Sigma Y)^2]}} \qquad (6.13)$$

6.6 Linear Relationships

For the paired (X,Y) values given in Table 6.1 we found that $r = 1.00$. In Figure 6.2 we have plotted these paired (X,Y) values and we see that the plotted points fall directly on a straight line. If we plot any other set of (X,Y) values and if the plotted points fall directly on a straight line, with nonzero slope, the correlation coefficient between the X and Y values will be equal to either 1.00 or -1.00. If the slope of the straight line is upward from left to right, as in Figure 6.2, the correlation coefficient will be equal to 1.00. If the slope of the straight line is downward from left to right, the correlation coefficient will be equal to -1.00. For example, if we were to plot the paired (X,Y) values given in Table 6.2, and for which we found that $r = -1.00$, we would see that these plotted points also fall directly on a straight line that slopes downward from left to right.

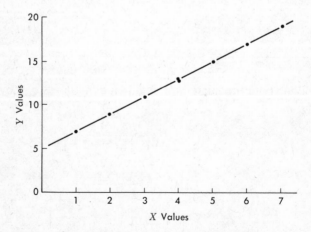

Figure **6.2** — Plot of the X and Y values given in Table 6.1.

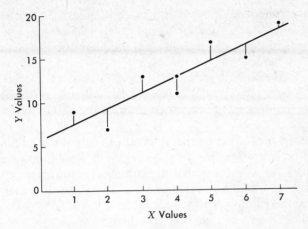

In general, if we have paired measurements on two variables, X and Y, for a sample of n subjects, we will not find the correlation coefficient to be equal to either 1.00 or -1.00. In other words, the plotted points will not all fall directly on a straight line. For example, in Figure 6.3 we have plotted the paired (X,Y) values given in Table 6.4. For these paired (X,Y) values we see that it is impossible to draw a straight line in such a way that all of the points fall on it. The reason why this is so is because, in this instance, we have

$$r = \frac{50}{\sqrt{(28)(112)}} = \frac{50}{58} = 0.89$$

instead of $r = 1.00$ or $r = -1.00$.

Table **6.4** — Calculation of the Residual Sum of Squares $\Sigma(Y - \tilde{Y})^2$

(1) Obser-vation	(2) X	(3) Y	(4) x	(5) y	(6) x^2	(7) y^2	(8) xy	(9) $\tilde{Y}$	(10) $Y - \tilde{Y}$	(11) $(Y - \tilde{Y})^2$
1	7	19	3	6	9	36	18	18.4	0.6	0.36
2	6	15	2	2	4	4	4	16.6	−1.6	2.56
3	5	17	1	4	1	16	4	14.8	2.2	4.84
4	4	13	0	0	0	0	0	13.0	0.0	0.00
5	4	11	0	−2	0	4	0	13.0	−2.0	4.00
6	3	13	−1	0	1	0	0	11.2	1.8	3.24
7	2	7	−2	−6	4	36	12	9.4	−2.4	5.76
8	1	9	−3	−4	9	16	12	7.6	1.4	1.96
Σ	32	104	0	0	28	112	50		0.0	22.72

Whenever the *general trend* of a set of plotted (X,Y) points is either upward or downward, from left to right, and when the points appear to follow a straight line, even though all of the points do not fall on the line, X and Y may be described as having *some degree* of *linear relationship*. For example, the plotted points in Figure 6.3 indicate that there is a linear relationship between X and Y, even though only one of the eight points falls on the line that has been drawn through the points to represent the general trend.

It is necessary that the general trend of the plotted points be either downward or upward in order for a linear relationship to exist between two variables. This is to say that the straight line drawn to represent the trend must have either a positive or a negative slope. For example, if we were to plot the paired (X,Y) values $(1,5)$, $(2,5)$, $(3,5)$, and $(4,5)$, these points would fall on a straight line. But this straight line would be parallel to the X axis; that is, the straight line would have zero slope and there is no linear relationship between these paired (X,Y) values.

In the example described above, it is obvious that the Y values have zero variance; that is, $s_Y^2 = 0$. In Section 6.3 we showed that the covariance

$$c_{XY} = \frac{\Sigma(X - \bar{X})(Y - \bar{Y})}{n-1}$$

could also be expressed as

$$c_{XY} = rs_X s_Y$$

It is clear that if either s_X^2 or s_Y^2 is equal to zero, the corresponding standard deviation, s_X or s_Y, will also be equal to zero. In this case both c_{XY} and r will be equal to zero. It is possible, however, for r to be equal to zero, even though neither s_X nor s_Y is equal to zero. For example, if $c_{XY} = 0$, then r must be equal to zero, regardless of the values of s_X and s_Y. Consequently, the most general statement we can make is that if $c_{XY} = 0$, there can be no linear relationship between X and Y. In other words, whenever $c_{XY} = 0$, it will also be true that $r = 0$.

We consider an example in which $s_X \neq 0$ and $s_Y \neq 0$. Assume we have the following paired (X,Y) values: $(-2,4)$, $(-1,1)$, $(0,0)$, $(1,1)$, and $(2,4)$. If the Y values are plotted against the X values, it will be seen that the trend of the points is at first downward and then upward. In this example, neither the X nor the Y values have zero variance, but the covariance is equal to zero. There is no linear relationship between these paired values; that is, $r = 0$. We emphasize that to say there is no linear relationship between two variables, does not mean that the variables are unrelated. In this example, the Y values are directly related to the X values by the simple equation $Y = X^2$, but this is not a *linear relationship*.

The correlation coefficient is a measure of the degree of linear relationship between two variables, X and Y. If there is no linear relationship, in the sense described above, the correlation coefficient will be equal to zero.

6.7 The Regression of Y on X

The line that we have drawn in Figure 6.3 is called a *regression line*[1] of Y on X. The regression line is not simply any straight line, but one that is fitted to the paired (X,Y) values so as to minimize the sum of the squared deviations of the plotted points from the line.

The equation of the regression line is given by

$$\tilde{Y} = \bar{Y} + b_Y(X - \bar{X}) \tag{6.14}$$

and this equation is called a *regression equation*. The value of b_Y is obtained by finding

$$b_Y = \frac{\Sigma(X - \bar{X})(Y - \bar{Y})}{\Sigma(X - \bar{X})^2} = \frac{\Sigma xy}{\Sigma x^2} \tag{6.15}$$

and b_Y is called the *regression coefficient* of Y on X.

For the data of Table 6.4, we have $\Sigma xy = 50$ and $\Sigma x^2 = 28$. Then, for the value of the regression coefficient, we have

$$b_Y = \frac{50}{28} = 1.8$$

We also have $\Sigma Y = 104$ and $\bar{Y} = 104/8 = 13$, and $\Sigma X = 32$ and $\bar{X} = 32/8 = 4$. Then the equation of the line drawn in Figure 6.3 is

$$\tilde{Y} = 13 + (1.8)(X - 4)$$

6.8 The Standard Error of Estimate

The value of $\tilde{Y}$, as given by the regression equation, is a *predicted value* of Y for any corresponding value of X. The discrepancy between Y and $\tilde{Y}$ is, in this sense, an error of prediction. The values of $\tilde{Y}$ corresponding to each value of X in Table 6.4 are given in column (9) of the table. These values were obtained by using the regression equation

$$\tilde{Y} = 13 + (1.8)(X - 4)$$

---

[1]Francis Galton used the term "regression" in studying the inheritance of stature. It was his observation that, on the average, the offspring of abnormally tall parents and abnormally short parents tend to move back toward the population mean height. The line describing this trend was called a "regression line." The term is still used to describe the line drawn among a group of points to represent the trend, but it no longer necessarily carries the original implications that Galton intended.

For example, for the observation in the first row, we have $X = 7$. Then, for this observation we have

$$\tilde{Y} = 13 + (1.8)(7-4) = 18.4$$

and the error of prediction is

$$Y - \tilde{Y} = 19.0 - 18.4 = 0.6$$

The errors of prediction for each of the other values of Y were obtained in the same manner and are given in column (10) of Table 6.4. We see that the sum of the errors of prediction, $\Sigma(Y - \tilde{Y})$, the sum of column (10), is equal to zero. The squares of the errors of prediction are given in column (11) of the table, and we find that

$$\Sigma(Y - \tilde{Y})^2 = 22.72$$

This sum of squares is less than the sum of squared errors we would obtain if we substituted any other value of b_Y in the regression equation such that this new value is not equal to $\Sigma xy / \Sigma x^2$. The straight line that we have fitted to the points in Figure 6.3, in other words, minimizes the sum of the squared deviations from the line. These deviations are indicated by the vertical lines in Figure 6.3. If we divide the sum of squared errors of prediction by $n-2$, and take the square root, we obtain a statistic called the *standard error of estimate*. Thus

$$s_{Y \cdot X} = \sqrt{\frac{\Sigma(Y - \tilde{Y})^2}{n-2}} \qquad (6.16)$$

For the data of Table 6.4, we have $\Sigma(Y - \tilde{Y})^2 = 22.72$ and, therefore,

$$s_{Y \cdot X} = \sqrt{\frac{22.72}{8-2}} = 1.9$$

The standard error of estimate, $s_{Y \cdot X}$, is similar to the standard deviation, s_Y. But, whereas s_Y measures the variation of the Y values in terms of their deviations from a constant value (the mean, $\bar{Y}$), $s_{Y \cdot X}$ measures the variation of the Y values in terms of their deviations from their individually predicted values, $\tilde{Y}$, on the regression line. If X and Y are, in fact, linearly related, that is, if $r \neq 0$, then we might expect that, in general, the individual deviations, $Y - \tilde{Y}$, will tend to be smaller than the corresponding deviations $Y - \bar{Y}$. As a consequence, $\Sigma(Y - \tilde{Y})^2$ should be smaller than $\Sigma(Y - \bar{Y})^2$ and $s_{Y \cdot X}$ should be smaller than s_Y. This is, in fact, the case. In the example under discussion, we have $s_{Y \cdot X} = 1.9$, whereas

$$s_Y = \sqrt{\frac{\Sigma(Y - \bar{Y})^2}{n-1}} = \sqrt{\frac{\Sigma y^2}{n-1}} = \sqrt{\frac{112}{8-1}} = 4.0$$

and $s_{Y \cdot X} = 1.9$ is considerably smaller than $s_Y = 4.0$.

When the correlation coefficient is equal to -1.00 or 1.00, then the standard error of estimate will be equal to zero. As the correlation coefficient tends toward zero, then $s_{Y \cdot x}^2$ approaches the value of s_Y^2. It will not equal s_Y^2 because of the difference in the denominators. The relationship between these two statistics is such that

$$s_{Y \cdot x}^2 = \frac{n-1}{n-2} s_Y^2 (1 - r^2) \tag{6.17}$$

If we multiply both sides of (6.17) by $n - 2$, then we have

$$\Sigma (Y - \tilde{Y})^2 = \Sigma (Y - \bar{Y})^2 (1 - r^2)$$

$$= \Sigma (Y - \bar{Y})^2 - r^2 \Sigma (Y - \bar{Y})^2$$

$$= \Sigma (Y - \bar{Y})^2 - \frac{[\Sigma (X - \bar{X})(Y - \bar{Y})]^2}{\Sigma (X - \bar{X})^2}$$

or

$$\Sigma (Y - \tilde{Y})^2 = \Sigma y^2 - \frac{(\Sigma xy)^2}{\Sigma x^2} \tag{6.18}$$

From Table 6.4, we see that

$$\Sigma y^2 = \Sigma (Y - \bar{Y})^2 = 112$$

$$\Sigma x^2 = \Sigma (X - \bar{X})^2 = \ \ 28$$

and

$$\Sigma xy = \Sigma (X - \bar{X})(Y - \bar{Y}) = 50$$

Substituting in (6.18) with these values, we have

$$\Sigma (Y - \tilde{Y})^2 = 112 - \frac{(50)^2}{28} = 22.71$$

which is, within rounding errors, equal to the sum of column (11) in the table.

We see, therefore, that it is not necessary to obtain the individual values of $Y - \tilde{Y}$ in order to calculate $s_{Y \cdot x}$ because we also have

$$s_{Y \cdot x} = \sqrt{\frac{1}{n-2} \left[\Sigma y^2 - \frac{(\Sigma xy)^2}{\Sigma x^2} \right]} = \sqrt{\frac{\Sigma (Y - \tilde{Y})^2}{n-2}}$$

6.9 The Regression of X on Y

We have discussed the correlation coefficient as a measure of the linear relationship between two variables. It should be obvious, from an examination of the formula for the correlation coefficient, that $r_{XY} = r_{YX}$ and we have, therefore, dropped the subscripts. In our discussion of regression, however, we were concerned with the prediction of Y values from values of X. It should be clear that the regression line, regression coefficient, and standard error of estimate, as presented in the discussion, all referred to predicting Y values from X and not to predicting X values from Y. In general, in prediction problems, we are concerned with the prediction of one variable from knowledge of another variable. The variable we use as a basis for prediction we usually designate as the X variable, and the variable we are attempting to predict is designated the Y variable.

If, for some reason or another, we were also interested in predicting X values from Y, as well as predicting Y values from X, then, for the regression equation, we would have

$$\tilde{X} = \bar{X} + b_X(Y - \bar{Y}) \qquad (6.19)$$

where b_X is defined as

$$b_X = \frac{\Sigma(X - \bar{X})(Y - \bar{Y})}{\Sigma(Y - \bar{Y})^2} = \frac{\Sigma xy}{\Sigma y^2} \qquad (6.20)$$

For the data of Table 6.4, we have

$$\bar{X} = 32/8 = 4$$

$$\bar{Y} = 104/8 = 13$$

and

$$b_X = \frac{50}{112} = 0.45$$

The regression equation for predicting X values from Y would then be

$$\tilde{X} = 4 + 0.45(Y - 13)$$

The standard error of estimate, $s_{X \cdot Y}$, would be given by

$$s_{X \cdot Y} = \sqrt{\frac{\Sigma(X - \tilde{X})^2}{n - 2}} \qquad (6.21)$$

and the sum of squares of the errors of estimate by

$$\Sigma(X - \tilde{X})^2 = \Sigma(X - \bar{X})^2 - \frac{[\Sigma(X - \bar{X})(Y - \bar{Y})]^2}{\Sigma(Y - \bar{Y})^2}$$

or

$$\Sigma(X - \tilde{X})^2 = \Sigma x^2 - \frac{(\Sigma xy)^2}{\Sigma y^2} \qquad (6.22)$$

For the data of Table 6.4, we have

$$\Sigma(X - \tilde{X})^2 = 28 - \frac{(50)^2}{112} = 5.68$$

with

$$s_{X \cdot Y} = \sqrt{\frac{5.68}{8 - 2}} = 0.97$$

We see that $s_{X \cdot Y} = 0.97$ is considerably smaller than

$$s_X = \sqrt{\frac{\Sigma(X-\bar{X})^2}{n-1}} = \sqrt{\frac{\Sigma x^2}{n-1}} = \sqrt{\frac{28}{8-1}} = 2.00$$

just as $s_{Y \cdot X} = 1.9$ was considerably smaller than $s_Y = 4.0$.

6.10 The Two Regression Lines

In Figure 6.4, we have plotted both the regression line of Y on X and the regression line of X on Y for the data of Table 6.4. The regression line of Y on X indicates the predicted change in Y as X varies. We observe from this line, for example, that with each change in X of 1 unit, there is a corresponding predicted change in Y of 1.8 units, and this is the value of the regression coefficient b_Y. Similarly, if we consider the regression line of X on Y, we find that with each change in Y of 1 unit, there is a corresponding predicted change in X of 0.45 units, and this is the value of the regression coefficient b_X. As the correlation coefficient approaches either of its two limiting values, -1.00 or 1.00, the two regression lines move closer together until, in the limiting case, with r equal to either -1.00 or 1.00, we have a single regression line, as we would have for the data of Table 6.1 or Table 6.2.

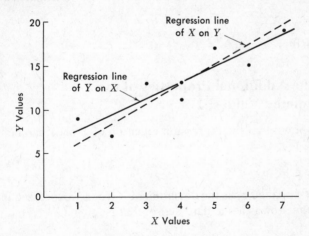

Figure **6.4**—Plot of the X and Y values of Table 6.4 showing the regression line of Y on X and the regression line of X on Y.

6.11 The Special Case of Regression When $s_X = s_Y$

We have defined the regression coefficient of Y on X as

$$b_Y = \frac{\Sigma(X-\bar{X})(Y-\bar{Y})}{\Sigma(X-\bar{X})^2}$$

If we divide both the numerator and denominator of the expression on the right by $n-1$, the numerator will be the covariance and the denominator will be the variance of the X measures and, in this instance, we have

$$b_Y = \frac{c_{XY}}{s_X^2}$$

Early in the chapter we showed that $c_{XY} = rs_X s_Y$. Substituting with this identity in the above expression, we also have

$$b_Y = \frac{rs_X s_Y}{s_X^2}$$

or

$$b_Y = r\frac{s_Y}{s_X} \tag{6.23}$$

In the same manner, we could show that the regression coefficient of X on Y is given by

$$b_X = r\frac{s_X}{s_Y} \tag{6.24}$$

In the special case where the standard deviation of the X measures is equal to the standard deviation of the Y measures, we would also have

$$b_Y = b_X = r$$

because, in this case, $s_Y/s_X = s_X/s_Y = 1$.

6.12 Some Additional Properties of the Predicted Values: $\tilde{Y}$

We have shown that the regression equation of Y on X can be written as

$$\tilde{Y} = \bar{Y} + r\frac{s_Y}{s_X}(X - \bar{X})$$

But we have defined z_X as the deviation of X from $\bar{X}$ expressed in standard deviation units or as a standard score, that is, $z_X = (X - \bar{X})/s_X$. Then we also have

$$\tilde{Y} = \bar{Y} + rs_Y z_X \tag{6.25}$$

If we subtract $\bar{Y}$ from both sides of (6.25) and then divide both sides by s_Y, we have

$$\frac{\tilde{Y} - \bar{Y}}{s_Y} = rz_X \tag{6.26}$$

In this instance, the left side of (6.26) is the predicted deviation of $\tilde{Y}$ from $\bar{Y}$ in terms of standard deviation units of the Y variable.

It is interesting to note that unless $r = 1.00$ or $r = -1.00$, the deviation of $\tilde{Y}$ from $\bar{Y}$ in terms of standard deviation units of the Y variable cannot be as great as the deviation of X from $\bar{X}$ in standard deviation units of the X variable. For example, if $r = 0.80$ and if X is two standard deviations above $\bar{X}$, then $rz_X = (0.80)(2.00) = 1.60$, and $\tilde{Y}$ will be only 1.60 standard deviations above the mean of the Y variable. Similarly, if $r = -0.80$ and if X is two standard deviations *above* $\bar{X}$, then $rz_X = (-0.80)(2.00) = -1.60$ and $\tilde{Y}$ will be only 1.60 standard deviations *below* the mean of the Y variable.

It is obvious from (6.25) that the mean of the predicted $\tilde{Y}$ values is equal to $\bar{Y}$, because

$$\Sigma \tilde{Y} = n\bar{Y} + rs_Y \Sigma z_X$$

and, as we stated previously, the mean of a set of standard scores is equal to zero; that is, $\Sigma z_X / n = 0$. Therefore,

$$\frac{\Sigma \tilde{Y}}{n} = \bar{Y}$$

If we subtract $\bar{Y}$ from both sides of (6.25), then we have

$$\tilde{Y} - \bar{Y} = rs_Y z_X$$

and because $\bar{Y}$ is equal to the mean of the $\tilde{Y}$ values, the variance of the predicted values will be given by

$$s_{\tilde{Y}}^2 = \frac{\Sigma(\tilde{Y} - \bar{Y})^2}{n-1} = \frac{r^2 s_Y^2 \Sigma z_X^2}{n-1}$$

But we have also stated that for the variance of a set of standard scores, we have

$$s_{z_X}^2 = \frac{\Sigma(z_X - \bar{z}_X)^2}{n-1} = \frac{(\Sigma z_X - 0)^2}{n-1} = \frac{\Sigma z_X^2}{n-1} = 1$$

and, therefore,

$$s_{\tilde{Y}}^2 = r^2 s_Y^2$$

We see that unless $r = 1.00$ or $r = -1.00$, the variance of the predicted values $(\tilde{Y})$ must always be smaller than the variance of the observed values (Y). Only if $r = 1.00$ or $r = -1.00$ can we have

$$s_{\tilde{Y}}^2 = s_Y^2$$

REVIEW OF TERMS

bivariate distribution
correlation coefficient
covariance
deviation product sum
linear relationship

negative relationship
positive relationship
predicted value
regression coefficient
regression equation

regression line   standard score
standard error of estimate

REVIEW OF SYMBOLS

r
$\bar{X}$ z
Σxy b_Y
$s_{Y \cdot X}$ b_X
$s_{X \cdot Y}$ $\tilde{Y}$

QUESTIONS AND PROBLEMS

6.1   We have the following paired values of X and Y:

X: 16   14   13   10   8   7   5   4   2   1
Y: 5   3   2   8   9   17   6   15   11   14

Calculate each of the following statistics:

(a) r (d) b_Y
(b) $\bar{X}$ (e) b_X
(c) $\bar{Y}$ (f) $s_{Y \cdot X}$

6.2   Using the regression equation for predicting Y values from X values, find the values of $\tilde{Y}$ for each of the following values of X: (a) $X = 2$, (b) $X = 8$, and (c) $X = 14$, in Problem 6.1.

6.3   Using the regression equation for predicting X values from Y values, find the values of $\tilde{X}$ for each of the following values of Y: (a) $Y = 5$, (b) $Y = 9$, and (c) $Y = 14$, in Problem 6.1.

6.4   Make a plot of the X and Y values for Problem 6.1. Using the predicted values obtained in Problems 6.2 and 6.3, it is possible to draw the regression line of Y on X and the regression line of X on Y. Draw these two lines and label each one.

6.5   If we find $\tilde{Y}$ for a value of X equal to the mean of the X distribution, what is the predicted value?

6.6   If we find $\tilde{X}$ for a value of Y equal to the mean of the Y distribution, what is the predicted value?

6.7   Do both regression lines pass through the point $(\bar{X}, \bar{Y})$? Examine the two regression equations and explain why this is so.

6.8   What does the regression coefficient b_Y measure? What would it mean, for example, if b_Y were equal to 0.75?

6.9   What does the standard error of estimate measure? What would it mean, for example, if $s_{Y \cdot X}$ were equal to 6.0 and s_Y were equal to 10.0?

6.10   Under what condition can b_Y be equal to b_X?

6.11 John has a standard score of $z_X = (X - \bar{X})/s_X = 1.5$ on X. The correlation between X and Y is equal to 0.80. By how much will his predicted score, $\tilde{Y}$, deviate from the mean, $\bar{Y}$, in terms of standard deviation units? In other words, find $(\tilde{Y} - \bar{Y})/s_Y$.

6.12 If b_Y is equal to 0.15 and b_X is equal to 0.60, then what is the value of r?

6.13 If r is equal to 0.60, s_Y is equal to 8.0, and s_X is equal to 10.0, then what is b_X equal to?

6.14 If b_X is equal to 0.15, s_X is equal to 6.0, and r is equal to 0.40, then what is s_Y equal to?

6.15 Suppose we have a variable X with the following values: 1, 2, 3, 4, 4, 5, 6, and 7. We subtract $\bar{X}$ from each value of X and divide the resulting deviation by s; that is, we transform each value of X into standard scores z. (a) What will the mean of this distribution of standard scores be equal to? (b) What will the variance of this distribution of standard scores be equal to?

6.16 What statements can you make describing the plot of paired (X,Y) values, if you know that $r = -1.00$?

6.17 If r is equal to zero, then what can you say about the standard error of estimate?

6.18 If s_X is equal to 8.0 and s_Y is equal to 8.0, and r is equal to 0.80, what are the values of b_Y and b_X?

6.19 If r is equal to -1.00, then what can you say about the standard error of estimate?

6.20 A student states that in the physical sciences, where measures of variables are highly accurate, it is not unusual to find that the correlation between two variables is 1.00 or higher. Do you agree with his statement? Why?

6.21 Each of the sets of X values shown below has a standard deviation equal to 1.58. The Y values paired with the X values in each set were obtained by multiplying the X values by a constant b and then adding another constant a to the product. (a) Find the values of a and b for each of the sets of Y values. (b) Are the values of b that you obtained equal to the regression coefficient b_Y? Explain why.

(a)		(b)		(c)		(d)		(e)	
X	Y	X	Y	X	Y	X	Y	X	Y
1	3	4	12	3	3	5	−9	5	5
3	5	3	9	6	9	1	−1	6	7
5	7	2	6	7	11	3	−5	7	9
2	4	6	18	4	5	4	−7	3	1
4	6	5	15	5	7	2	−3	4	3

6.22 We have the following paired values of X and Y:

X	Y
2	1
3	2
1	3

(a) What is the value of the covariance? (b) What is the value of the correlation coefficient?

6.23 For the paired values given below, calculate the correlation coefficient.

X	Y
4	4
5	3
6	5
3	1
2	2

6.24 We have n values of a variable X and for each X we let

$$Y = a + bX$$

where a and b are constants. (a) What is the value of the correlation coefficient between X and Y? Explain why. (b) What is the value of the regression coefficient of Y on X? Explain why. (c) What is the value of $s_{Y \cdot X}$? Explain why.

6.25 The covariance of X and Y for a given set of n paired values is equal to 10.0. To each of the Y values we add a constant a to obtain $Z = Y + a$. What is the value of the covariance of X and Z? Explain why.

6.26 The correlation coefficient between X and Y is equal to 0.80. If we add a constant to each of the Y values to obtain $Z = Y + a$, will the correlation coefficient between X and Z be equal to 0.80? Explain why.

6.27 The correlation coefficient between X and Y is equal to 0.60. We multiply each value of Y by a constant b to obtain $Z = bY$. Will the correlation coefficient between X and Z be equal to 0.60? Explain why.

6.28 We have the paired X and Y values shown below:

X	Y
2	6
4	4
6	2

(a) Explain why, for this example, the correlation coefficient between X and Y must be equal to -1.00. (b) For the same example, explain why s_X must be equal to s_Y. (c) For this example, what do you know about the value of the covariance?

6.29 If $z = (X - \bar{X})/s$, can you prove that: (a) $\bar{z} = 0$ and that (b) $s_z = 1$?

6.30 For a sample of n observations, we have $\bar{X} = 10$ and $s = 2$. We change each value of X in the various ways shown below to obtain a new variable Y.
(a) $Y = 2X$
(b) $Y = X - 2$
(c) $Y = X/2$
(d) $Y = X + 2$
(e) $Y = (X - 10)/2$

Explain why the correlation coefficient between the paired X and Y values for each of the above must be equal to 1.00.

6.31 Find the correlation coefficient for the following paired values of X and Y:

X	Y
5	3
3	5
4	4
1	1
2	2

7

Elementary Probability Theory

7.1 Introduction

The statistics we have studied so far are useful in describing the particular set of observations we have made. Often, however, the questions we desire to answer in terms of a set of observations go beyond the problem of description. If we have available the mean of a sample, we may wish to use this mean as a basis for inferring something about the mean of the population from which the sample was drawn. In other cases, we may assume that a sample was drawn from a specified population, and we wish to determine whether this is a reasonable assumption. In still other cases, we may have two or more sets of observations of the same variable but with each set obtained under different conditions. We wish to determine, on the basis of statistics we have calculated describing these sets of observations, whether or not it is reasonable to conclude that these samples are homogeneous, that is, whether they are from the same population or not.

To understand how statistical techniques can be used in answering questions such as the ones we have raised above requires that we have some understanding of elementary probability theory. Almost any college algebra text contains a chapter on probability, and the subject is one that you have encountered in one form or another previously. In order to make the discussion readily understandable to those who have not had college algebra or who have forgotten their brief encounter with the subject, the examples presented in this chapter are quite simple.

In a sense, it is unfortunate that these simple examples, useful in discussing probability, depart from subject-matter interests. The drawing of a ball from a box would appear to have little relationship to the observation of whether a rat turns right or left in a maze or to a subject's response in a discrimination experiment. And yet the observations made by drawing balls from a box provide us with simple models of the behavior of rats in a maze or the behavior of subjects in discrimination experiments. We can only emphasize that it is not the example that is of importance. Rather, the importance attaches to the principle we hope to illustrate by means of the example. The statistical techniques we shall learn to apply in dealing with the rat's response in a maze or the student's response in a discrimination experiment are based on the same principles we derive from observing balls drawn from a box.

7.2 Sample Spaces and Probabilities

Let us suppose that we have a set of six table tennis balls. We number these balls 1, 2, 3, 4, 5, and 6. The balls are then placed in a box that has an opening just sufficient in size to permit a single ball to pass through. We shake the box thoroughly and turn it upside down, permitting one ball to drop out, and we record the number written on the ball. We may regard the drawing of a ball from the box as a simple experiment in which the possible outcomes of the experiment consist of one of the numbers 1, 2, 3, 4, 5, or 6.

The set of all possible outcomes of an experiment is called a *sample space* for the experiment. We indicate a sample space by the letter S. The possible outcomes of the experiment are the elements of the set S and are called *sample points*. We use braces to enclose the elements of S and, in the example under discussion, we have

$$S = \{1, 2, 3, 4, 5, 6\} \tag{7.1}$$

The ordering of the sample points within the braces is unimportant. What is important is that if S is to be the sample space for an experiment, each possible outcome of the experiment must correspond to one and only one element or sample point.

To each sample point in S we wish to assign a number P called a *probability*. If the numbers assigned to the sample points are to be called probabilities, then they must satisfy the following conditions:

1. The values of P assigned to the sample points must be equal to or greater than zero and equal to or less than 1, that is, $0 \leqslant P \leqslant 1$.
2. The sum of the values of P assigned to the sample points must be equal to 1; that is, we must have $\Sigma P = 1$.
3. If we divide the sample points in S into subsets such that the subsets are mutually exclusive and exhaustive, then the values of P associated with each subset must also sum to 1.

In the experiment under consideration, we shall assume that the process we have described of selecting a ball from the box results in each ball having an equal chance of being selected. Thus, to each of the sample points in S as defined by (7.1), we assign a value of $P = 1/6$. Now, suppose that on one of the balls we write the letter A, on two of the balls we write the letter B, and on the remaining three balls the letter C. We now draw a ball from the box in the manner described and record the letter that appears on the ball. As a sample space for this experiment, we have

$$S = \{A, B, C\} \tag{7.2}$$

Because only one of the balls has a letter A on it, we shall assume that $P(A) = 1/6$. We can obtain a B by drawing either one of the two balls that have the letter B and we shall assume, therefore, that $P(B) = 2/6$. Because three of the balls have the letter C, we shall assume that $P(C) = 3/6$. Note that the values of P assigned to the sample points of S, as defined by (7.2), satisfy the probability axioms.

7.2.1 INDEPENDENT EVENTS: THE MULTIPLICATION RULE

Suppose that after we have drawn one ball from the box and recorded the letter appearing on it, we replace the ball in the box, shake it thoroughly, and draw a second ball from the box. If we make two observations in the manner specified, what is the probability that the first observation will be A and the second B? The probability of A on the first draw is $1/6$. If we replace the ball in the box, the probability of B on the second draw is $2/6$. The probability that we will draw an A followed by a B is the product of the two probabilities or $(1/6)(2/6) = 2/36$. In this instance, we say that the two events A and B are *independent*, because the probability of obtaining B on the second draw is in no way conditional on whether or not A occurred on the first draw. If two or more events are independent, the probability that they will all occur is the product of their separate probabilities. We shall refer to this statement as the *multiplication rule*. The procedure of replacing each ball in the box after it is drawn ensures that the probabilities of obtaining an A, B, or C on each of the successive draws will remain the same, regardless of what has happened on the previous draw. Thus the outcomes of these draws are independent.

7.2.2 MUTUALLY EXCLUSIVE EVENTS: THE ADDITION RULE

What is the probability that on a single draw the observation will be either a B or a C? The probability of drawing a B is $2/6$, and the probability of drawing a C is $3/6$. The probability of drawing either a B or a C will be given by the sum of these two probabilities. Thus

$$P(B \cup C) = P(B) + P(C) \tag{7.3}$$

where $P(B \cup C)$ is read "the probability of *either* a B *or* a C." In our example, we have

$$P(B \cup C) = 2/6 + 3/6 = 5/6$$

Mutually exclusive events are those in which if one of the events happens, none of the others can occur. In our example, on a single draw, we may obtain either a B or a C, but we cannot obtain both. These two events are, therefore, mutually exclusive. In general, the probability that any one of a number of mutually exclusive events will occur is equal to the sum of the probabilities of the separate events. We shall refer to this statement as the *addition rule*.

7.3 Conditional Probability

Suppose we now modify the sampling procedure by *not replacing* a ball in the box after it is drawn. On the first draw the probabilities of obtaining A, B, or C remain as originally stated. But we shall now see that the probabilities on the second draw are changed, depending on the nature of the first draw. If we make two observations, what is the probability that the second ball will be a B, subject to the condition that the first ball was an A? The probability of obtaining A on the first draw is 1/6. If we do not replace the A in the box, we shall then have only five balls in the box, of which two are B's and three are C's. We shall assume that any one of the five balls has a probability of 1/5 of being selected. The probability of drawing a B on the second draw, subject to the condition that we have drawn an A on the first draw, will, therefore, be 2/5. We refer to this probability as a conditional probability. In *conditional probability*, we are concerned with the probability of an event subject to the condition that other events have already occurred. To indicate a conditional probability, we shall use the notation $P(B|A)$ which we read as "the probability of B given that A has occurred."

We can use the notion of conditional probability to define independent and nonindependent events. Two events, A and B, are said to be independent if the probability of B occurring is in no way influenced by the fact that A either has occurred or has not occurred. If the probability of B is changed by either the occurrence or nonoccurrence of A, then the two events are not independent. *If events are not independent, the probability that a given sequence of events will occur is the product of their conditional probabilities.*

If we draw three balls from the box without replacement after each of the draws, what is the probability that the three observations will occur in the order A, B, and C? We have $P(A) = 1/6$ on the first draw. We have found that $P(B|A) = 2/5$. If both A and B have occurred, we will have four balls left in the box, of which three will be C's. Then, for the conditional probability of C, given that both A and B have occurred, we have $P(C|A,B) = 3/4$. Then the probability of obtaining A, B, C, in the order specified, will be

$$P(A, B, C) = P(A)P(B|A)P(C|A, B)$$
$$= (1/6)(2/5)(3/4)$$
$$= 1/20$$

7.4 Permutations

Suppose that we have three distinct objects, A, B, and C. A given order or sequence or arrangement of these objects is called a *permutation*. We see that there are six different orders or permutations possible for these three objects. They are

$$\begin{array}{ccc} ABC & BAC & CAB \\ ACB & BCA & CBA \end{array}$$

The number of different permutations of n objects is always equal to $n!$ This symbol is called *factorial n*. Factorial n, or $n!$, means the product of all of the integers from n to 1. We indicate the number of permutations of n objects taken n at a time by $_nP_n$. Thus

$$_nP_n = n! \tag{7.4}$$

With three objects, the number of permutations is $3! = 3 \times 2 \times 1 = 6$, and these permutations are shown above for the objects we have called A, B, and C.

How many different permutations of A, B, and C can we have if we consider only two objects at a time? These possible different orders are as follows:

$$\begin{array}{ccc} AB & BC & AC \\ BA & CB & CA \end{array}$$

The number of different permutations of n objects taken r at a time is given by[1]

$$_nP_r = \frac{n!}{(n-r)!} \tag{7.5}$$

For three objects taken two at a time, we have

$$_3P_2 = \frac{3!}{(3-2)!} = 6$$

different orders, as shown above.

Let us assume that we have n objects, and that these objects can be divided into k sets so that the objects within each set are alike. We let r_1,

_____
[1] We may have $(n-r)! = 0!$. We always take $0! = 1$.

$r_2, \ldots, r_k$ represent the number of objects within each of the respective sets, with $n = r_1 + r_2 + \cdots + r_k$. Then the number of permutations of the n objects will be given by

$$_nP_{r_1, r_2, \ldots, r_k} = \frac{n!}{r_1! r_2! \ldots r_k!} \tag{7.6}$$

where $_nP_{r_1, r_2, \ldots, r_k}$ is the number of permutations of n objects of which r_1 are alike, r_2 are alike, and so on. For example, if we have two A's, one B, and one C, then the number of permutations of these four objects would be

$$_4P_{2,1,1} = \frac{4!}{2!1!1!} = 12$$

The twelve permutations of these four objects are given below:

AABC	ABAC	ABCA	BAAC	BACA	BCAA
AACB	ACAB	ACBA	CAAB	CABA	CBAA

We note that if we have only two sets of objects with the objects within each set alike, then if one set has r objects, the second will have $n-r$ objects. Under this circumstance, we can write (7.6) as

$$_nP_{r,(n-r)} = \frac{n!}{r!(n-r)!} \tag{7.7}$$

Thus, with four objects such that two are A's and two are B's we would have

$$_4P_{2,2} = \frac{4!}{2!(4-2)!} = 6$$

permutations. These six permutations are given below:

AABB	ABAB	ABBA
BBAA	BABA	BAAB

7.5 Combinations

The number of different ways in which we can select r objects from a set of n with the order or arrangement of the r objects ignored is called the number of *combinations* of the n objects taken r at a time. The number of combinations of n objects taken r at a time is given by[2]

$$_nC_r = \frac{n!}{r!(n-r)!} \tag{7.8}$$

and we note that this formula is the same as (7.7).

---

[2]In other books, the number of combinations of n things taken r at a time is indicated by $\binom{n}{r}$.

If we have three objects, A, B, and C, then we have already seen that taking these objects two at a time, we have six permutations: $AB, BA, BC, CB, AC,$ and CA. Ignoring the order, AB and BA are one combination, BC and CB another combination, and AC and CA a third combination. Thus, in this instance, we have only three combinations. This result can be obtained directly from (7.8). With $n = 3$ and $r = 2$, we have, by substitution in (7.8),

$$_3C_2 = \frac{3!}{2!(3-2)!} = 3$$

combinations.

7.6 Random Samples

Let us consider a *population* of three observations. We designate the observations by means of three balls on which we have placed the letters A, B, and C, respectively. The balls are placed in a box such as the one described previously. The method of sampling is to shake the box and then to turn it upside down, permitting one ball to drop out. We then shake the box again, *without replacing* the first ball, and draw a second observation. These two observations will comprise a sample. We are interested in the nature of the possible samples drawn in this way. If we list the observations in the order drawn, the different samples we could obtain would be:

$$AB \quad AC \quad BC$$
$$BA \quad CA \quad CB$$

What is the probability of obtaining the sample AB? The probability of obtaining A on the first draw is $1/3$ and the conditional probability of B on the second draw, given that A has occurred, is $1/2$. The probability of obtaining the sample AB is, therefore, $(1/3)(1/2) = 1/6$. This is also the probability of obtaining each of the other samples. Thus, in this instance, each set of $n = 2$ observations has the same probability of being the sample drawn.

What is the probability that A will be an observation included in the sample? We note that four out of the six samples include A, and the probability that the sample will include A is, therefore, $4/6$ or $2/3$. This is also the probability that the sample will include B and the probability that it will include C.

We have thus seen that every possible sample of $n = 2$ observations has the same probability of being drawn and that each single observation has the same probability of being included in the sample, if we can assume that the method of sampling is random. These properties are often used to define a simple random sample or, more briefly, a *random sample*. The use of the term "random" in connection with a sample of observations should, however, be considered as applying to the particular procedure or method of obtaining the observations rather than to the sample itself. A random sample of observations, in other words, is one obtained by a particular method that we believe provides randomness in the observations selected. We shall describe more useful procedures of *random selection* later.

Note also that the term "random selection" applies to the observations and not to the values of the observations. In all of the statistical techniques that we shall learn to apply to the values of the observations in order to infer something that goes beyond the mere description of the observations, we shall have to invoke the notion of randomness. If we are merely interested in describing a sample set of observations in terms, say, of the mean and standard deviation, we need specify nothing more about the nature of the observations other than that we hope we have not made errors in recording the observations or in the arithmetic used in calculating the mean and standard deviation. The concept of randomness enters in when we wish to use statistical techniques in answering questions that go beyond a description of the obvious characteristics of the sample.

REVIEW OF TERMS

addition rule for probabilities
combination
conditional probability
factorial n
independent events
multiplication rule for
   probabilities

permutation
probability axioms
random sample
random selection
sample point
sample space

REVIEW OF SYMBOLS

$_nP_n$ $n!$ $_nP_r$

$_nP_{r,(n-r)}$ $_nC_r$ $_nP_{r_1,r_2,\ldots,r_k}$

$P(A)$ $P(B|A)$ $P(A|B)$

QUESTIONS AND PROBLEMS

7.1   Find the value of each of the following:

(a) $6!$

(b) $\dfrac{6!}{(6-3)!}$

(c) $\dfrac{6!}{3!(6-3)!}$

(d) $\dfrac{6!}{2!(6-2)!}$

(e) $\dfrac{6!}{3!2!1!}$

7.2   A box contains four balls with A's, two with B's, and one with C marked on them. If a single ball is drawn at random from the box, give the probability that it will be:

(a) A (d) A or B
(b) B (e) A or C
(c) C (f) B or C

7.3   If, from the same box, the first ball drawn is replaced and a second ball is drawn, give the probability of obtaining, in the order indicated:

(a) AB (d) BC
(b) AC (e) CA
(c) BA (f) CB

7.4   Under the same conditions of sampling as described in Problem 7.3, give the probability that the sample will contain:

(a) $2A$'s   (d) at least $1A$ (1 or more)
(b) $2B$'s   (e) at least $1B$ (1 or more)
(c) $2C$'s   (f) at least $1C$ (1 or more)

7.5   On a multiple-choice test of four items, each item with four alternatives, what is the probability of getting exactly three items correct by chance?

7.6   In a discrimination experiment, the probability of a correct discrimination is assumed to be $P = 1/3$. A student is given $n = 5$ trials and it is assumed that the probability of a correct discrimination remains the same on each trial. We let T be the number of correct discriminations in the $n = 5$ trials. Find the number of ways in which each of the following values of T can be obtained:

(a) $T = 5$ (d) $T = 2$
(b) $T = 4$ (e) $T = 1$
(c) $T = 3$ (f) $T = 0$

7.7   For the experiment described in Problem 7.6, find the probability of each of the following values of T:

(a) $T = 5$ (d) $T = 2$
(b) $T = 4$ (e) $T = 1$
(c) $T = 3$ (f) $T = 0$

7.8   A student claims that he did not know any of the answers to a true-false test consisting of six items. Assume, therefore, that the probability of a correct response to each item is $P = 1/2$. (a) What is the probability that he will have answered all six items correctly? (b) What is the probability that he will have answered five items correctly? (c) What is the probability that he will have answered four or more items correctly?

7.9   An instructor teaches two courses, A and B, and a seminar C. He has 10 students in A, 15 in B, and 5 in C. Five of the students enrolled in A are also enrolled in B. The names of the students are placed in a box and one name is selected at random. Give the probability that the student whose name is drawn will be enrolled in:

(a) *A*

(b) *B*

(c) *C*

(d) both *A* and *B*

(e) *A* or *B*

(f) Given that the name of a student enrolled in *B* has been drawn, what is the probability that he will also be enrolled in *A*?

(g) Given that the name of a student enrolled in *A* has been drawn, what is the probability that he will also be enrolled in *B*?

7.10 A committee of three students must be selected from a group of 10. No one volunteers, and the instructor places the names of the 10 students in a box and selects three names at random. How many different samples of three can be drawn? We are not interested in the order in which the names are drawn.

7.11 We have six males and four females. We wish to draw a sample of five such that the sample includes three males and two females. How many different samples of the kind described can be drawn? We are not interested in the order in which the sample is drawn.

7.12 We have six cardboard squares, each of a different color. The squares are to be presented to subjects, one at a time, and they are to be asked to express whether they like or dislike each color. We feel that the order of presentation in which the squares are presented may be of importance, and we decide, therefore, to present the colors in all possible orders. We also decide to test two subjects with each of the possible orders. How many subjects will we have to test?

7.13 A battery of 15 achievement tests is given to a group of students. If we were to find the intercorrelations between all possible pairs of the tests, how many correlation coefficients would we have to calculate?

7.14 We have used the notation $P(A)$ to refer to the probability of A, and $P(A|B)$ to refer to the probability of A, given that B has occurred. Consider a set of 25 disks such that on each one we have marked one of the pairs of letters AC, AD, BC, or BD. The table at the left shows the pairs of letters and the table at the right gives the corresponding frequency for each pair.

	C	D			C	D	Total
A	AC	AD		A	7	3	10
B	BC	BD		B	5	10	15
				Total	12	13	25

Assume that the disks are placed in a box and that one is drawn at random. Give the probability for each of the following:

(a) $P(A)$

(b) $P(B)$

(c) $P(A|C)$

(d) $P(B|C)$

(e) $P(A|D)$ (i) $P(C|A)$
(f) $P(B|D)$ (j) $P(D|A)$
(g) $P(C)$ (k) $P(C|B)$
(h) $P(D)$ (l) $P(D|B)$

Now note that the probability of obtaining a disk with A and D is

$$P(A \text{ and } D) = P(A) \times P(D|A) = P(D) \times P(A|D) = 3/25$$

(m) Substitute in the above expression with the appropriate probabilities. Note also that the probability of obtaining a disk with A or D is

$$P(A \text{ or } D) = P(A) + P(D) - P(A \text{ and } D) = 20/25$$

(n) Substitute in the above expression with the appropriate probabilities. On the basis of the above discussion, find the following probabilities:

(o) $P(B \text{ and } D)$ (r) $P(A \text{ or } C)$
(p) $P(B \text{ or } D)$ (s) $P(B \text{ and } C)$
(q) $P(A \text{ and } C)$ (t) $P(B \text{ or } C)$

7.15 Fifteen subjects are to be assigned to one of three groups in such a way that each group has five subjects. In how many ways is this possible? We are not interested in the order in which the subjects are assigned.

7.16 Five students meet in a seminar with their instructor. There are five chairs which the students may occupy. They are determined that they will not seat themselves in exactly the same way at any meeting of the seminar. For how many meetings can they do this before duplicating a previous seating arrangement?

7.17 I place five objects on my desk and concentrate on two of the objects. You try to read my mind. Assuming that you cannot and that any answer you give is a matter of chance, what is the probability that you will guess the correct two objects?

7.18 State, in words, the meaning of each of the following:

(a) $n!$ (c) $_nP_r$
(b) $_nC_r$ (d) $_nP_{r_1, r_2, \ldots, r_k}$

7.19 What is the essential difference between the number of permutations of n objects taken r at a time and the number of combinations of n objects taken r at a time?

7.20 If we know that two events are mutually exclusive, can we also conclude that they are independent? Explain the basis of your answer.

7.21 A battery of ten tests is given to a group of students. If we were to find the intercorrelations between all possible pairs of tests, how many *different* correlation coefficients would we have to calculate?

7.22 A cage contains five rats. Three of the rats are males and two are females. An experimenter selects a rat at random from the cage and puts it aside without observing its sex. A second rat is then selected from the four remaining rats in the cage. (a) What is the probability that the second rat selected is a male? (b) What is the probability that the second rat selected is a female? (c) What is the probability that both the first and second rats selected are males? (d) What is the probability that both the first and second rats selected are females? (e) What is the probability that at least one of the three rats remaining in the cage, after the second rat has been selected, is a female?

7.23 From a group consisting of five students, three of whom are females and two of whom are males, a random sample of $n = 3$ is selected without replacement. (a) What is the probability that the sample will consist of three females? (b) What is the probability that the sample will consist of two females and one male? (c) What is the probability that the sample will consist of one female and two males? (d) What is the probability that the sample will consist of three males?

7.24 One permutation of the four letters $A, B, C,$ and $D,$ is $DBCA.$ What is the total number of ways in which we can permute these four letters?

7.25 If a random sample of $n = 3$ is selected from the eight letters $A, B, C, D, E, F, G,$ and $H,$ without replacement, how many different ordered samples are possible?

7.26 In how many different ways can we classify nine students into three groups of three students each? We assume that the order in which the students are put into each group is unimportant.

7.27 A random sample of $n = 4$ is selected without replacement from a cage containing six male rats and four female rats. (a) In how many ways is it possible to obtain a sample of two females and two males? (b) In how many ways is it possible to obtain a sample of three males and one female? (c) In how many ways is it possible to obtain a sample of four males? (d) In how many ways is it possible to obtain a sample of four females?

8

Random Sampling Distributions

8.1 Introduction

If you have a clear understanding of the principles developed in the previous chapter, then you should be prepared to apply them to the cases described in this chapter. Let us suppose, for example, that we have a population such that the only possible values of a variable X are 5, 4, 3, 2, and 1. We shall further assume that each of the five values occurs equally frequently in the population so that if an observation is drawn at random from the population, then each of the five possible values of X has the same probability, $P = F/N = 0.2$, of occurring. We have chosen this particular population only because it permits us to illustrate in a very simple manner the principles in which we are interested.

8.2 The Mean and Variance of a Population

In column (3) of Table 8.1 we give the products of PX, that is, the product of a given value of X times its corresponding probability. The sum of the entries in this column is 3.0, and this sum is equal to

$$\mu = \Sigma PX \tag{8.1}$$

where we have used μ to indicate the population mean. The population mean, therefore, is equal to 3.0.

In column (4) of the table we have given the deviations of $X - \mu$ and in column (5) the squares of these deviations. Column (6) shows the products of $P(X - \mu)^2$. The sum of the entries in this column is 2.0, and this sum is equal to

$$\sigma^2 = \Sigma P(X - \mu)^2 \tag{8.2}$$

Table **8.1** — Calculation of the Mean and Variance for a Population with a Known Distribution

(1) X	(2) P	(3) PX	(4) $X - \mu$	(5) $(X - \mu)^2$	(6) $P(X - \mu)^2$
5	0.2	1.0	2	4	0.8
4	0.2	0.8	1	1	0.2
3	0.2	0.6	0	0	0.0
2	0.2	0.4	−1	1	0.2
1	0.2	0.2	−2	4	0.8
Σ	1.0	3.0	0		2.0

where we have used σ^2 to indicate the population variance. The population variance, therefore, is equal to 2.0.

The mean and variance just obtained are independent of N, the number of observations in the population. The only requirement we have made is that the observations be distributed uniformly over the five possible values of the variable, that is, so that $P = 0.2$ for each value. As long as this requirement is satisfied, the population mean will be equal to 3.0, and the variance will be equal to 2.0.

8.3 Random Samples of $n = 2$ Observations Drawn from the Population

We are interested in the *sampling distribution*, also called the *probability distribution*, of various statistics based on random samples of size n drawn from the population under consideration. If we draw an observation at random from the population, the probability of obtaining an observation with any one of the five possible values would be $P = 0.2$. If the population is indefinitely large, the conditional probabilities relating to the value of a second observation could, for all practical purposes, be considered to be equal to the original probabilities relating to the first draw.

We can simulate this population and ensure that the probabilities associated with each of the five possible values remain the same in the following way: We put each value of the variable on a ball, place the five balls in a box, and thoroughly mix them. We then draw one ball from the box, record the value, and replace it before making the second draw. We shall assume that this method of sampling will provide us with a random sample of $n = 2$ observations.

As a sample space for the outcomes of this experiment, we note that any one of the five possible values may occur on the first draw and, because any one of the five possible values may also occur on the second draw, we have a total of $5 \times 5 = 25$ pairs of *ordered* values. Table 8.2 shows the possible values that the two observations may take, if the samples are drawn in the manner described. The first number in the cell

Table **8.2** — Values of Observations of Samples of $n = 2$ Drawn from the Population of Table 8.1. The First Entry in a Cell of the Table Is the Value of the First Draw, and the Second Entry Is the Value of the Second Draw.

Value of First Draw	Value of Second Draw				
	5	4	3	2	1
5	5,5	5,4	5,3	5,2	5,1
4	4,5	4,4	4,3	4,2	4,1
3	3,5	3,4	3,3	3,2	3,1
2	2,5	2,4	2,3	2,2	2,1
1	1,5	1,4	1,3	1,2	1,1

refers to the value of the first observation and the second number to the value of the second observation. Each of these 25 ordered pairs of values has the same probability of 1/25 of being drawn, if the method of sampling is random. Table 8.3 gives the sums, means, and differences, respectively, for each of the 25 samples.

Table **8.3** — Distribution of Sums, Means, and Differences for Samples of $n = 2$ Drawn from the Population of Table 8.1. The First Entry in a Cell of the Table Is the Sum, the Second Entry the Mean, and the Third the Difference for Each Possible Sample.

Value of First Draw	Value of Second Draw				
	5	4	3	2	1
5	10	9	8	7	6
	5.0	4.5	4.0	3.5	3.0
	0	1	2	3	4
4	9	8	7	6	5
	4.5	4.0	3.5	3.0	2.5
	−1	0	1	2	3
3	8	7	6	5	4
	4.0	3.5	3.0	2.5	2.0
	−2	−1	0	1	2
2	7	6	5	4	3
	3.5	3.0	2.5	2.0	1.5
	−3	−2	−1	0	1
1	6	5	4	3	2
	3.0	2.5	2.0	1.5	1.0
	−4	−3	−2	−1	0

8.4 Probability Distribution of the Mean

From the means given in Table 8.3, we form the frequency distribution shown in Table 8.4. We note that there is only one way in which we can obtain a sample with a mean of 5.0 (5,5), and the probability of a sample with this mean is $1/25 = 0.04$. There are two samples that yield a mean of 4.5 (5,4 and 4,5). Because the probability of each of these samples is 0.04, and because they are mutually exclusive, the probability of a sample mean of 4.5 is $0.04 + 0.04 = 0.08$. Similarly, we see that the probability of a sample mean of 3.5 is 0.16, and so on. The probabilities corresponding to all possible values of $\bar{X}$ are given in column (2) of Table 8.4. The distribution shown in Table 8.4 is called a *sampling distribution* or a *probability distribution*.

Table **8.4** — Calculation of the Mean and Variance for the Distribution of Means Given in Table 8.3

(1) $\bar{X}$	(2) P	(3) $P\bar{X}$	(4) $\bar{X} - \mu$	(5) $(\bar{X} - \mu)^2$	(6) $P(\bar{X} - \mu)^2$
5.0	0.04	0.20	2.0	4.00	0.16
4.5	0.08	0.36	1.5	2.25	0.18
4.0	0.12	0.48	1.0	1.00	0.12
3.5	0.16	0.56	0.5	0.25	0.04
3.0	0.20	0.60	0.0	0.00	0.00
2.5	0.16	0.40	−0.5	0.25	0.04
2.0	0.12	0.24	−1.0	1.00	0.12
1.5	0.08	0.12	−1.5	2.25	0.18
1.0	0.04	0.04	−2.0	4.00	0.16
Σ	1.00	3.00	0.0		1.00

We show the calculation of the mean of the probability distribution in column (3) of Table 8.4. The entries in this column are the values of the sample means multiplied by the corresponding probabilities. The sum of column (3) is equal to $\Sigma P\bar{X}$ and is, therefore, the mean of the sample means. We note that the mean of the distribution is 3.0 and that this is exactly equal to the population mean of the original values of X. We can say that the mean of a random sample of size n is an *unbiased* estimate of the population mean, because its average value is equal to the population mean. We note also that the distribution of the sample means is symmetrical about the population mean, as shown in Figure 8.1.

8.5 The Variance and Standard Error of the Mean

In column (4) of Table 8.4, we show the deviation of each possible value of the sample mean from the population mean. The entries in this column

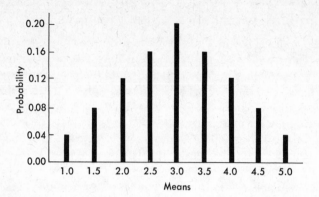

Figure **8.1** — A graphic representation of the distribution of means shown in Table 8.4. The population from which the samples of $n = 2$ observations were drawn is given in Table 8.1.

are values of $\bar{X} - \mu$. The squares of these deviations are given in column (5). Multiplying the entries in column (5) by the corresponding probabilities given in column (2), we obtain the products shown in column (6). The sum of this column is 1.00 and is equal to $\Sigma P(\bar{X} - \mu)^2$ or the *variance* of the means. Thus, in general,

$$\sigma_{\bar{X}}^2 = \Sigma P(\bar{X} - \mu)^2 \tag{8.3}$$

We note that this variance is also given by

$$\sigma_{\bar{X}}^2 = \frac{\sigma^2}{n} \tag{8.4}$$

where $\sigma_{\bar{X}}^2$ = the variance of the mean
  σ^2 = the known population variance of the individual values of X
  n = the sample size

Because we have already shown, in Table 8.1, that the population variance is equal to 2.0, and because we have $n = 2$ as the sample size, we also have by substitution in (8.4)

$$\sigma_{\bar{X}}^2 = \frac{2}{2} = 1.00$$

Formula (8.4) is a general formula and is not limited to the particular population we have specified. In general, if we have a large number of samples of n independent and randomly drawn observations from a population with variance σ^2, the variance of the means of such samples will be equal to the population variance divided by n. Taking the square root of (8.4), we obtain the *standard error* of the mean. Thus

$$\sigma_{\bar{X}} = \frac{\sigma}{\sqrt{n}} \tag{8.5}$$

where $\sigma_{\bar{X}}$ = the standard error of the mean
$\quad\quad \sigma$ = the population standard deviation
$\quad\quad n$ = the sample size

When the population variance is not known but is estimated from a random sample of n observations, we substitute the estimate of the population variance, as given by $s^2 = \Sigma(X - \bar{X})^2/(n-1)$, in (8.4). The estimated variance of the mean will then be

$$s_{\bar{X}}^2 = \frac{s^2}{n} \quad\quad\quad (8.6)$$

and the estimated standard error of the mean will be

$$s_{\bar{X}} = \frac{s}{\sqrt{n}} \quad\quad\quad (8.7)$$

8.6 The Mean and Variance of a Difference

Table 8.5 gives the distribution of the differences, $D = X_1 - X_2$, for each of the possible samples of $n = 2$ observations. The mean of these differences, as shown by the sum of column (3), is equal to zero, and this is the average difference between randomly paired independent observations drawn from the same population. The sum of the entries in column (6) is equal to the variance of the differences, or

$$\sigma_{X_1-X_2}^2 = \Sigma P(D - \mu_D)^2 \quad\quad\quad (8.8)$$

and, for the example under consideration, we have $\sigma_{X_1-X_2}^2 = 4.0$.

The variance of the differences will also be given by

$$\sigma_{X_1-X_2}^2 = \sigma_{X_1}^2 + \sigma_{X_2}^2 \quad\quad\quad (8.9)$$

Table **8.5** — Calculation of the Mean and Variance for the Distribution of Differences Given in Table 8.3

(1) $D = X_1 - X_2$	(2) P	(3) PD	(4) $D - \mu_D$	(5) $(D - \mu_D)^2$	(6) $P(D - \mu_D)^2$
4	0.04	0.16	4	16	0.64
3	0.08	0.24	3	9	0.72
2	0.12	0.24	2	4	0.48
1	0.16	0.16	1	1	0.16
0	0.20	0.00	0	0	0.00
−1	0.16	−0.16	−1	1	0.16
−2	0.12	−0.24	−2	4	0.48
−3	0.08	−0.24	−3	9	0.72
−4	0.04	−0.16	−4	16	0.64
Σ	1.00	.00	0		4.00

where $\sigma_{X_1-X_2}{}^2$ = the variance of the differences, $X_1 - X_2$

$\sigma_{X_1}{}^2$ = the variance of the values of X_1

$\sigma_{X_2}{}^2$ = the variance of the values of X_2

The values of X_1, we know, were drawn from a population with known variance of $\sigma^2 = 2.0$, as were also the values of X_2. The subscripts merely serve to identify the first and second observation in each sample. Then, according to (8.9), we have

$$\sigma_{X_1-X_2}{}^2 = 2.0 + 2.0 = 4.0$$

and this is the same value we obtained by direct calculation in Table 8.5.

The variance of the difference between pairs of *independent* random observations drawn from any two populations will always be equal to the sum of the variances of the two populations. If the paired observations are drawn from identical populations, as in the case at hand, or from populations with identical variances so that $_{X_1}{}^2 = \sigma_{X_2}{}^2$, then

$$\sigma_{X_1-X_2}{}^2 = 2\sigma^2 \tag{8.10}$$

where σ^2 is the common population variance.

The standard error of the distribution of differences will be given by the square root of (8.9) and will be equal to

$$\sigma_{X_1-X_2} = \sqrt{\sigma_{X_1}{}^2 + \sigma_{X_2}{}^2} \tag{8.11}$$

It should be noted that (8.9) and (8.11) are general formulas for *independent* observations and are not limited to the particular population under consideration.

If we sample from populations with unknown variances which are estimated by $s^2 = \Sigma (X - \bar{X})^2/(n-1)$, then the formula for the variance of the differences becomes

$$s_{X_1-X_2}{}^2 = s_{X_1}{}^2 + s_{X_2}{}^2 \tag{8.12}$$

and the standard error becomes

$$s_{X_1-X_2} = \sqrt{s_{X_1}{}^2 + s_{X_2}{}^2} \tag{8.13}$$

8.7 The Mean and Variance of a Sum

In Table 8.6 we show the calculation of the mean and variance of the sums T. We see that the mean of the sums, as given by the sum of column (3), is equal to 6.0, and the variance of the sums, as given by the sum of column (6), is 4.0. In general, if we have n randomly drawn observations from the same population or from populations with the same mean, the expected or average value of the sums will be given by

$$\mu_T = n\mu \tag{8.14}$$

Table **8.6** — Calculation of the Mean and Variance for the Distribution of Sums Given in Table 8.3

(1) $T = X_1 + X_2$	(2) P	(3) PT	(4) $T - \mu_T$	(5) $(T - \mu_T)^2$	(6) $P(T - \mu_T)^2$
10	0.04	0.40	4	16	0.64
9	0.08	0.72	3	9	0.72
8	0.12	0.96	2	4	0.48
7	0.16	1.12	1	1	0.16
6	0.20	1.20	0	0	0.00
5	0.16	0.80	−1	1	0.16
4	0.12	0.48	−2	4	0.48
3	0.08	0.24	−3	9	0.72
2	0.04	0.08	−4	16	0.64
Σ	1.00	6.00	0		4.00

where μ_T = the expected or average value of the sums
      n = the number of observations contributing to the sum
      μ = the common population mean

Substituting in (8.14) with $n = 2$ and with $\mu = 3.0$, we have $\mu_T = 2 \times 3.0 = 6.0$, the same value we obtained in Table 8.6 by direct calculation.

A general formula for the variance of a sum T of n independent randomly drawn observations from n populations is

$$\sigma_T^2 = \sigma_{X_1}^2 + \sigma_{X_2}^2 + \cdots + \sigma_{X_n}^2 \qquad (8.15)$$

where the successive terms on the right refer to the respective variances of the observations in each of the n populations. If the observations are drawn from the same population, as in the case at hand, or from populations with the same population variance, then (8.15) can be written

$$\sigma_T^2 = n\sigma^2 \qquad (8.16)$$

where σ^2 refers to the common population variance and n refers to the number of observations contributing to the sum. Because we have, in the example under consideration, $n = 2$ and $\sigma^2 = 2.0$, (8.16) gives 4.0 as the variance of the sums, and this is the same value we obtained by direct calculation in Table 8.6.

If the common population variance is not known, but is estimated by $s^2 = \Sigma(X - \bar{X})^2/(n - 1)$, then ns^2 will provide an estimate of $n\sigma^2$.

8.8 Distribution of the Variance

In Table 8.7 we show the distribution of the variances for the samples of $n = 2$ observations drawn from the population given in Table 8.1. The obtained values of X_1 and X_2 for each possible sample of $n = 2$

Table 8.7 — Calculation of the Mean of the Distribution of Variances for the Samples of Table 8.3

(1) s^2	(2) P	(3) Ps^2
8.0	0.08	0.64
4.5	0.16	0.72
2.0	0.24	0.48
0.5	0.32	0.16
0.0	0.20	0.00
Σ	1.00	2.00

observations are given in Table 8.2. For each of these 25 samples we could calculate the sample variance

$$s^2 = \frac{\Sigma(X - \bar{X})^2}{n-1}$$

However, as we prove in answer to one of the problems at the end of the chapter, there is a much simpler method for obtaining the sample variance for the special case of $n = 2$ observations. For any sample with only $n = 2$ observations, it will be true that

$$s^2 = \frac{(X_1 - X_2)^2}{2}$$

The values of $X_1 - X_2$ for each of the 25 samples shown in Table 8.2 are given in Table 8.3. We see, in Table 8.3, that we have five samples for which $(X_1 - X_2)^2 = 0$ and for each of these samples we have $s^2 = 0$. Because each of these five samples has a probability of $1/25 = 0.04$ of occurring and because they are mutually exclusive, the probability of obtaining $s^2 = 0$ will be equal to $(5)(0.04) = 0.20$ and this probability has been entered in column (2) of Table 8.7 opposite $s^2 = 0$. Similarly, we have eight samples in Table 8.3 for which $(X_1 - X_2)^2 = 1$, and for each of these samples we will have $s^2 = 1/2 = 0.50$. Each of these eight samples has the same probability, $1/25 = 0.04$, of occurring and because they are mutually exclusive, the probability of $s^2 = 0.50$ will be equal to $(8)(0.04) = 0.32$ and this probability has been entered in column (2) of Table 8.7 opposite $s^2 = 0.50$. The other entries in columns (1) and (2) of Table 8.7 were obtained in a similar manner.

The products of the corresponding entries in columns (1) and (2) are given in column (3). We note that the sum of column (3) gives the average value of the sample variances and is equal to 2.0. This is exactly equal to the value of the population variance σ^2, as we showed earlier. Thus, we can say that the sample variance, as defined by $s^2 = \Sigma(X - \bar{X})^2/(n-1)$, is an *unbiased* estimate of the population variance σ^2, because the long run

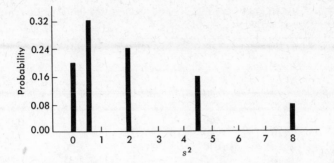

Figure **8.2** — A graphic representation of the distribution of variances shown in Table **8.7**. The population from which the samples of $n = 2$ observations were drawn is given in Table **8.1**.

average value of s^2 will be equal to σ^2 provided we have random samples drawn from a common population.

We observe, however, that the distribution of the sample variances, shown in Table 8.7 and Figure 8.2, is not symmetrical about the population variance, but is right skewed. It can be shown, however, that as the sample size n increases, the distribution of the sample variances does become more symmetrical about the population variance.

8.9 Sampling from a Finite Population

In drawing the samples of $n = 2$, we replaced the first ball before drawing the second. If we had not, then it would not have been possible to have obtained samples in which the values of the variable were the same, for example, 5 and 5. That is because we simulated the population by placing only five balls in the box, with only one ball corresponding to each possible value. By replacing the ball in the box after the first draw, we could simulate an infinite population in which the probabilities of obtaining each possible value of the variable remained the same on each draw.

If there are only five balls in the box, one for each of the five possible values of X, and if the first ball is not replaced before drawing the second ball, then it would be, as pointed out above, impossible to obtain a sample in which the two values of the variable were the same. In this case, there are only 20 possible samples, each with a probability of $1/20 = 0.05$. The distribution of the sample means, for $n = 2$ observations drawn without replacement from five balls, is shown in Table 8.8. We note that the mean of the distribution is 3.0, and this is the value of the population mean. However, the variance of the means, as shown by the sum of column (6), is no longer equal to 1.00, the value we obtained in Table 8.4 or by (8.4), but instead is only 0.75.

When we sample without replacement from a finite population of N

Table **8.8** — Calculation of the Mean and Variance for the Distribution of Means Obtained in Sampling without Replacement

(1) $\bar{X}$	(2) P	(3) $P\bar{X}$	(4) $\bar{X} - \mu$	(5) $(\bar{X} - \mu)^2$	(6) $P(\bar{X} - \mu)^2$
4.5	0.10	0.45	1.5	2.25	0.225
4.0	0.10	0.40	1.0	1.00	0.100
3.5	0.20	0.70	0.5	0.25	0.050
3.0	0.20	0.60	0.0	0.00	0.000
2.5	0.20	0.50	−0.5	0.25	0.050
2.0	0.10	0.20	−1.0	1.00	0.100
1.5	0.10	0.15	−1.5	2.25	0.225
Σ	1.00	3.00	0.0		0.750

observations with variance σ^2, the variance of the mean is given by

$$\sigma_{\bar{X}}^2 = \left(\frac{N-n}{N-1}\right)\frac{\sigma^2}{n} \qquad (8.17)$$

where $\sigma_{\bar{X}}^2 =$ the variance of the mean
$N =$ the number of observations in the population
$n =$ the sample size
$\sigma^2 =$ the known population variance

If the population variance is not known, we substitute the sample estimate s^2 for σ^2 in (8.17) and obtain $s_{\bar{X}}^2$ rather than $\sigma_{\bar{X}}^2$.

The factor $(N-n)/(N-1)$ is a correction for sampling from a finite population or, what amounts to the same thing, sampling from a limited population without replacement. Thus, if we had not replaced the first ball in the box before drawing the second, we would have been drawing a sample of $n = 2$ from a finite population of $N = 5$. Substituting in (8.17) with the values of n, N, and σ^2, we would have

$$\sigma_{\bar{X}}^2 = \left(\frac{5-2}{5-1}\right)\frac{2}{2} = 0.75$$

which is the value of the variance of the means obtained for the distribution shown in Table 8.8.

If we take the square root of 0.75, we obtain $\sigma_{\bar{X}} = \sqrt{0.75} = 0.87$, rather than the value of 1.00, as given by (8.5). Thus we see that both the variance and standard error of the mean tend to be overestimated when the correction factor is ignored. The error is somewhat greater in the case of the variance than in the case of the standard error.

Suppose, however, that our sampling box contained two observations with the value of 5, two with the value of 4, and so on. We would then have had a finite population of $N = 10$. If we draw a sample of $n = 2$ without replacement from this population, the variance of the mean, as given by

(8.17), would be

$$\sigma_{\bar{X}}^2 = \left(\frac{10-2}{10-1}\right)\frac{2}{2} = 0.89$$

and the standard error of the mean would be $\sigma_{\bar{X}} = \sqrt{0.89} = 0.94$, rather than 1.00, the value we would obtain by (8.5). We thus see, in this instance, if the sample includes only $n/N = 0.2$ of the observations in the population, we would not be seriously in error if we ignored the correction factor in calculating the standard error of the mean. In general, we shall assume that the sampling fraction, n/N, is 0.2 or less and we shall, therefore, ignore the correction factor in the standard error formulas developed later.

REVIEW OF TERMS

finite population correction factor   sample space
probability distribution   standard error of the mean
random sample

REVIEW OF SYMBOLS

ΣPX	$\sigma_{\bar{X}}^2$	$\sigma_{X_1-X_2}^2$
$\Sigma P(X-\mu)^2$	$s_{\bar{X}}^2$	D
μ_T	μ_D	$(N-n)/(N-1)$
σ_T^2	s^2	σ^2
$\Sigma P(\bar{X}-\mu)^2$	μ	$s_{X_1-X_2}^2$

QUESTIONS AND PROBLEMS

8.1 Assume that in a given population X can take only one of five possible values. These values of X are shown below and underneath each value is the probability associated with the value.

Value of X: 5   4   3   2   1
Probability: 0.1   0.2   0.4   0.2   0.1

(a) What is the value of the population mean μ? (b) What is the value of the population variance σ^2?

8.2 From the population described in Problem 8.1, we draw random samples of $n = 2$ with replacement after each draw. The probability of drawing a sample with X values of 5 and 5 will be $(0.1)(0.1) = 0.01$; the probability of drawing a sample of X values of 5 and 4, in that order, will be $(0.1)(0.2) = 0.02$. The probabilities associated with all possible ordered samples can be obtained by multiplying the corresponding probabilities of the two values of X. Show these probabilities by completing the table given below.

Value of First Draw		Value of Second Draw				
		5	4	3	2	1
	Probability	0.1	0.2	0.4	0.2	0.1
5	0.1	0.01	0.02			
4	0.2					
3	0.4			0.16		
2	0.2					
1	0.1					

8.3 A table consisting of the means, sums, and differences for the samples drawn in Problem 8.2 can be constructed. The cell entries in the table would be exactly the same as those in Table 8.3 in the text. The probabilities associated with these cell entries will be those that you calculated in Problem 8.2. (a) Make a distribution of the sample means and show the probability associated with each possible value of the mean. (b) What is the value of the mean of this distribution? (c) What is the value of the variance of the distribution of sample means? (d) What is the value of the standard error of the mean? It should be equal to $\sigma/\sqrt{n}$, where σ is the population standard deviation.

8.4 Suppose that a population standard deviation σ is equal to 10.0 Plot a curve showing the value of the standard error of the mean as n, the sample size, takes the following values: 4, 9, 16, 25, 36, 49, 64, 81, and 100.

8.5 Note that if $\sigma_{\bar{x}} = \sigma/\sqrt{n}$, and if we want to reduce the standard error by 1/2, then

$$\left(\frac{1}{2}\right)\sigma_{\bar{x}} = \left(\frac{1}{2}\right)\frac{\sigma}{\sqrt{n}} = \frac{\sigma}{\sqrt{4n}}$$

In other words, if for a given value of n we have a given value of $\sigma_{\bar{x}}$, and if we want to reduce this value of the standard error of the mean by 1/2, we must quadruple the sample size. If $\sigma_{\bar{x}} = 2.4$ for samples of $n = 30$, then what value of n would result in $\sigma_{\bar{x}} = 0.8$?

8.6 If the samples in Problem 8.2 had been drawn without replacement from a finite population with N equal to 10, then what would the standard error of the mean be equal to?

8.7 Assume that the samples of $n = 2$ in Problem 8.2 were drawn without replacement from a finite population. Make use of the finite population correction factor and give the standard error of the mean for the following values of N: (a) $N = 20$, (b) $N = 30$, and (c) $N = 50$.

8.8   We draw samples from a population with known variance σ^2 equal to 25.0 and mean μ equal to 15.0. In all cases we assume that the n observations are independently and randomly drawn and that the population is sufficiently large that we do not need to be concerned about a finite population correction factor. We draw an ordered sample of $n = 2$ and take the difference between the value of the first observation and the value of the second observation. (a) What value would we have for the mean of the distribution of differences? (b) What value would we have for the standard deviation of the distribution of differences?

8.9   For the population described in Problem 8.8, we draw random samples of $n = 3$ under the same conditions described. We find the sum of the three values of the three observations. (a) What value would we have for the mean of the distribution of sums? (b) What value would we have for the standard deviation of the distribution of sums?

8.10 For each of the samples of Problem 8.2, find the variance as defined by

$$s^2 = \frac{\Sigma (X - \bar{X})^2}{n - 1}$$

and where $\bar{X}$ is the mean of the sample and $n = 2$. (a) Multiply each value of s^2 by its corresponding probability. Is the average value thus obtained equal to the population variance? (b) Make a histogram showing the distribution of the values of s^2.

8.11 Use the data obtained in Problems 8.2 and 8.3 to make a distribution of the differences, $D = X_1 - X_2$. Multiply each value of D by its corresponding probability. (a) What is the mean value of the distribution of D? (b) What is the variance of the distribution of D? (c) Is the variance of D as calculated equal to $2\sigma^2$ where σ^2 is the population variance?

8.12 Use the data obtained in Problems 8.2 and 8.3 to make a distribution of the sums of the $n = 2$ values of X. Multiply each sum by its corresponding probability. (a) What is the mean of the distribution of sums? (b) What is the variance of the distribution of sums? (c) Is the variance of the distribution of sums equal to $2\sigma^2$, where σ^2 is the population variance?

8.13 We take an ace and a deuce from a deck of playing cards and place them face down in a random order on a table. We give a student another ace and deuce, and ask him to try to match his cards with those on the table. Assume that any placement he makes is a matter of chance. (a) What is the probability that he will make two correct matches? (b) What is the probability that he will make one correct match? Explain why. (c) What is the probability that he will make zero correct matches?

8.14 Let T be the number of correct matches in Problem 8.13. You have obtained the probability for each possible value of T. (a) What is the value of μ_T? (b) What is the value of σ_T^2?

8.15 Assume that we have exactly the same kind of problem as in Problem 8.13, but with three cards to be matched, the ace, deuce, and trey. Find the probability associated with each possible value of T. (a) What is the value of μ_T? (b) What is the value of σ_T^2?

8.16 Again, assume that we have the same kind of problem as in Problem 8.13, but with four cards to be matched, the ace, deuce, trey, and four. Find the probability associated with each possible value of T. (a) What is the value of μ_T? (b) What is the value of σ_T^2?

8.17 On the basis of the results obtained in the above problems, are you willing to generalize about the value of μ_T and σ_T^2? For example, what would you expect the values of μ_T and σ_T^2 to be, if there were $n = 10$ cards to be matched?

8.18 If $\sigma^2 = 9$ and $n = 4$, what is the value of $\sigma_{\bar{X}}$?

8.19 Assume X is a variable that can take the values of 2, 1, and 0 with corresponding probabilities of 0.1, 0.6, and 0.3. Find the values of σ and μ.

8.20 The standard error of a sample mean is $\sigma_{\bar{X}} = 3$, based on a sample of $n = 9$ observations. If we want to reduce the standard error by 1/4, how many observations should we include in the sample?

8.21 A fair die is rolled $n = 10$ times. Let T be the sum of the ten numbers face up. Calculate μ_T and σ_T^2.

8.22 If we have a sample with only $n = 2$ observations, we can designate the two values of X as X_1 and X_2. Can you prove that for this sample $s^2 = (X_1 - X_2)^2/2$?

9

Sampling from a Binomial Population

9.1 Introduction

In this chapter we consider a population in which the observations can take only the values of $X = 1$ or $X = 0$. A population of this kind is called a *binomial population*. We can simulate this population by placing in a box two sets of balls. On each ball in one set the number "1" is printed, and on each ball in the other set the number "0." We shake the box thoroughly and draw a single observation. We record whether the value of this observation is 0 or 1, and replace the ball in the box. We then take a second observation in the same manner, replacing the ball in the box after recording the value of the observation. We continue in this way until we have a sample of $n = 5$ observations. By replacing the ball in the box after each draw, we hold constant the probability of obtaining any given ball on each draw. If, for example, there are N balls in the box, the probability of obtaining any one of them on each draw is $1/N$. It is obvious, therefore, that every possible sample of n observations has the same probability of being drawn.

9.2 The Distribution of the Sum T

For any given sample of $n = 5$ observations, we have the possibility that 5, 4, 3, 2, 1, or 0 of the observations will have a value of 1 and these also are the possible values of $T = \Sigma X$. We are interested in the distribution of the sample values of T under the conditions specified. We shall assume that in the population represented by the N balls in the box, the proportion P of observations with a value of 1 is 0.5, and the proportion Q of observations with a value of 0 is also 0.5, because $Q = 1 - P$. Having

117

specified the sample size n and the population proportion P, we can now derive the distribution of the sample T's.

Because $P = 0.5$, we may say that, if our method of sampling is random, the probability of obtaining an observation with a value of $X = 1$ on each draw is $1/2$, and this is also the probability of obtaining an observation with a value of $X = 0$. These probabilities refer to the values of the variable. Because we have replaced each observation before taking another, these probabilities will remain unchanged on successive draws. By the multiplication rule for independent probabilities, we see that the probability of obtaining a sample in which the values of the observations, in order, are 11111 is

$$\frac{1}{2} \times \frac{1}{2} \times \frac{1}{2} \times \frac{1}{2} \times \frac{1}{2} = \frac{1}{32}$$

This is the only sample for which the sample value of T could be 5. The probability of obtaining a sample in which the values of the observations are, in order, 11110, is also $1/32$. This sample would be one in which the sample value of T is 4. But we observe that there are other possible samples that could result in $T = 4$. These would be samples in which we have, in order, 11101, 11011, 10111, and 01111. There are, in other words, five mutually exclusive ways in which we could obtain a sample in which $T = 4$. By the addition rule for the probability of any one of a set of mutually exclusive events, we see that the probability of obtaining a sample in which $T = 4$ is $5/32$.

In how many different ways could we obtain a sample for which $T = 3$? Because these would be samples in which we have three 1's and two 0's,

Table 9.1 — Samples of $n = 5$ Drawn from a Binomial Population with $P = 0.5$

Samples	Value of T	Samples	Value of T
11111	5	11000	2
11110	4	10100	2
11101	4	10010	2
11011	4	10001	2
10111	4	01100	2
01111	4	01010	2
11100	3	01001	2
11010	3	00110	2
11001	3	00101	2
10101	3	00011	2
10011	3	10000	1
10110	3	01000	1
01110	3	00100	1
01101	3	00010	1
01011	3	00001	1
00111	3	00000	0

we see that

$$_nP_{r,(n-r)} = \frac{n!}{r!(n-r)!}$$

will give the number of ways in which we can permute $r = 3$ values of $X = 1$ and $n - r = 2$ values of $X = 0$. Thus, there are

$$_5P_{3,2} = \frac{5!}{3!2!} = 10$$

ways in which we could obtain samples of three 1's and two 0's. These mutually exclusive ways are shown in Table 9.1, where we also show all of the other possible ordered samples.

9.3 The Binomial Distribution

Each of the samples listed in Table 9.1 has a probability of $1/32 = 0.03125$ of occurring. There is only one sample with $T = 5$ and, therefore, $P(T = 5) = 0.03125$. There are five samples for which $T = 4$ and these samples are mutually exclusive. The probability of obtaining $T = 4$ is, therefore, $(4)(0.03125) = 0.15625$. In the same manner we can obtain the probabilities corresponding to the other possible values of T and these are summarized in Table 9.2.

Table **9.2** — Sampling Distribution of T in Random Samples of $n = 5$ Drawn from a Binomial Population with $P = 0.5$

T	No. of Samples	P
5	1	0.03125
4	5	0.15625
3	10	0.31250
2	10	0.31250
1	5	0.15625
0	1	0.03125

The distribution shown in Table 9.2 is a probability distribution. It is also a special kind of probability distribution, known as the *binomial probability distribution*. The binomial distribution is obtained by expanding the binomial $(P+Q)^n$, where P is the probability that $X = 1$, $Q = 1 - P$ is the probability that $X = 0$, and n is the sample size. The distribution is given by

$$(P+Q)^n = P^n + nP^{n-1}Q + \frac{n(n-1)}{1 \times 2} P^{n-2}Q^2 + \frac{n(n-1)(n-2)}{1 \times 2 \times 3} P^{n-3}Q^3$$

$$+ \cdots + Q^n \tag{9.1}$$

If $P = 0.5$ and $n = 5$, then we have

$$(0.5+0.5)^5 = (0.5)^5 + (5)\,(0.5)^4(0.5) + (10)\,(0.5)^3(0.5)^2$$
$$+ (10)\,(0.5)^2(0.5)^3 + (5)\,(0.5)\,(0.5)^4 + (0.5)^5$$

and the successive terms give the probabilities for T equal to 5, 4, 3, 2, 1, and 0, respectively. Thus, if we ask what is the probability of obtaining a sample with $T = 3$, this is given by the third term in the binomial expansion above.

We note that the coefficients of P and Q in the binomial expansion give the number of ways in which we can obtain a sample with a specified number of 1's and 0's and that we can determine this directly by means of the formula for the number of permutations of n objects of which r are alike and $n-r$ are alike. The number of different ways (samples) in which we can permute three 1's and two 0's, for example, is given by

$$_5P_{3.2} = \frac{5!}{3!2!} = 10$$

For any one of these 10 samples, say 11100, the probability is P^3Q^2 and this probability is the same for all samples containing three 1's and two 0's. Thus

$$\frac{5!}{3!2!}P^3Q^2$$

is the probability of obtaining a sample with $T = 3$ and this is also the third term in the binomial expansion $(P+Q)^5$.

9.4 The Mean μ_T of the Distribution of T

In Table 9.3, we calculate the mean and standard deviation of the distribution of T, based on samples of $n = 5$ observations each drawn from a binomial population with $P = 0.5$ and $Q = 1 - P = 0.5$. The probabilities for each value of T are given in column (2) of the table. In column (3) of the table, we give the products of column (1) and column (2). If we sum

Table 9.3 — Computation of the Mean and Variance of T in Samples of $n = 5$ Drawn from a Binomial Population with $P = 0.5$

(1) T	(2) P	(3) PT	(4) $T - \mu_T$	(5) $(T - \mu_T)^2$	(6) $P(T - \mu_T)^2$
5	0.03125	0.15625	2.5	6.25	0.1953125
4	0.15625	0.62500	1.5	2.25	0.3515625
3	0.31250	0.93750	0.5	0.25	0.0781250
2	0.31250	0.62500	−0.5	0.25	0.0781250
1	0.15625	0.15625	−1.5	2.25	0.3515625
0	0.03125	0.00000	−2.5	6.25	0.1953125
Σ	1.00000	2.50000			1.2500000

these products we obtain the mean of the distribution or

$$\mu_T = \Sigma PT$$

and, in this example, we have

$$\mu_T = 2.5$$

We have previously shown that the mean of a binomial population is equal to P, the proportion of observations with $X = 1$ or, equivalently, the probability that $X = 1$ if an observation is selected at random from the population. We have also shown, in the previous chapter, that the mean of a sum of n observations drawn from the same population is simply $n\mu$, where μ is the population mean. In the present example, we have $\mu = P$ and, therefore,

$$\mu_T = nP \tag{9.2}$$

We have $n = 5$ and $P = 0.5$ in our example and, therefore,

$$\mu_T = (5)(0.5) = 2.5$$

which is the same as the sum of column (3) in Table 9.3.

9.5 The Variance σ_T^2 of the Distribution of T

In column (4) of Table 9.3 we give the values of $T - \mu_T$ and in column (5) the squares of these deviations. Multiplying each of the squared deviations by the corresponding probabilities, given in column (2), we obtain the products shown in column (6). The sum of these products is the variance for the distribution of T, that is,

$$\sigma_T^2 = \Sigma P(T - \mu_T)^2$$

and, in this example, we have

$$\sigma_T^2 = 1.25$$

We have previously shown that for a binomial population $\sigma^2 = PQ$. In the previous chapter we showed that if we have a sum of n independent observations drawn from a common population, then the variance of the sum is $n\sigma^2$, where σ^2 is the common population variance. For a binomial population, we have $\sigma^2 = PQ$ and, therefore,

$$\sigma_T^2 = nPQ \tag{9.3}$$

and

$$\sigma_T = \sqrt{nPQ} \tag{9.4}$$

For our example with $n = 5$, $P = 0.5$, and $Q = 0.5$, we have

$$\sigma_T^2 = (5)(0.5)(0.5) = 1.25$$

and

$$\sigma_T = \sqrt{(5)(0.5)(0.5)} = \sqrt{1.25} = 1.118$$

9.6 The Mean and Variance of the Mean of a Sample from a Binomial Population

If we have a sample of n observations from a binomial population, then $T = \Sigma X = f_1$, where f_1 is the number of observations in the sample with $X = 1$. The mean of this sample will be

$$\bar{X} = \frac{\Sigma X}{n} = \frac{f_1}{n} = p \tag{9.5}$$

and p is the proportion of observations in the sample with $X = 1$.

In the previous chapter we showed that the variance of a mean is given by

$$\sigma_{\bar{X}}^2 = \frac{\sigma^2}{n}$$

But for a binomial population, we have $\sigma^2 = PQ$ and, therefore,

$$\sigma_p^2 = \frac{PQ}{n} \tag{9.6}$$

and the standard error of p will be given by

$$\sigma_p = \sqrt{\frac{PQ}{n}} \tag{9.7}$$

If you will divide each of the values of T in column (1) of Table 9.3 by $n = 5$ to obtain the values of p and then multiply each value of p by the corresponding probabilities given in column (2), you will find that the average value of p is equal to $P = 0.5$. Similarly, if you subtract $P = 0.5$ from each of the sample values of p, square the resulting deviations and multiply the squared deviations by the probabilities given in column (2) of Table 9.3, the sum of these products will be the variance of the distribution of p. If your arithmetic is correct, then you will find that the sum of these products is equal to

$$\sigma_p^2 = \frac{(0.5)(0.5)}{5} = 0.05$$

9.7 Shape of the Distribution of T

Let us suppose that we have a binomial population with P equal to 0.5. Figure 9.1 shows the probability distribution of the sum T for random samples of $n = 5$ observations drawn from this population. The distribution of T is symmetrical about the average value $\mu_T = 2.5$, and this will

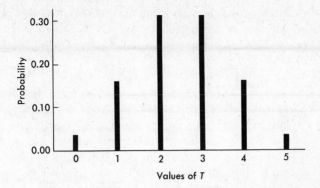

Figure **9.1** — The sampling distribution of the sum T for random samples of $n = 5$ drawn from a binomial population with $P = 0.5$.

always be true for random samples of size n drawn from a binomial population in which $P = Q$. For example, in Figure 9.2 we show the distribution of T for random samples of $n = 10$ observations drawn from a binomial population in which $P = 0.5$, and we note that this distribution is also symmetrical about the average value $\mu_T = 5.0$.

If n is small and if in the binomial population P is not equal to Q, then the distribution of T will not be symmetrical, but skewed. If P is greater than Q, the distribution will have a tail to the left; if P is less than Q, then the distribution of T will have a tail to the right. It can be shown, however, that as the sample size n is increased indefinitely, the distribution of T approaches, in the limit, a distribution that is symmetrical about the average value $\mu_T = nP$. As a guide, for practical purposes, we can assume that the

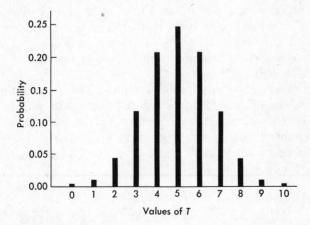

Figure **9.2** — The sampling distribution of the sum T for random samples of $n = 10$ drawn from a binomial population with $P = 0.5$.

distribution of T is approximately symmetrical about μ_T whenever both nP and nQ are equal to or greater than 5.

9.8 A Note to the Reader

The problems and exercises in this and the previous two chapters are based on probability models, and the answers given to these problems and exercises are based on the assumption that the probability model involved is a correct or true one. Your understanding of the material in these chapters will be increased if your curiosity motivates you to conduct experiments that will test empirically the reasonableness of the theorems and principles presented.

There are a number of ways in which you can conduct experiments involving various probability models. Tossing a set of n coins and counting the number of heads falling face up is a simple way of investigating the sampling distribution of T, a sum of n binomial variables. If the coins are unbiased, then the empirically derived distribution of T should not deviate greatly from the theoretical distribution described in the text.

A pair of unbiased dice, one red and one green, can be used to study the sampling distribution of sums, means, and differences. In addition, they can be used to study samples drawn from a binomial population with $P = 1/6$ (the probability of, say, a one face up), $P = 1/3$ (the probability of a one or two face up), and $P = 1/2$ (the probability of a one, two, or three face up).

A bridge deck of 52 cards can also be used to investigate empirically a variety of probability models. For example, the deck can be used to investigate samples drawn from a binomial population with $P = 1/2$ (the probability of drawing a red card) and $P = 1/4$ (the probability of drawing a card with a specified suit). If the cards of one suit are eliminated, then the remaining cards can be used to investigate samples from a binomial population with $P = 1/3$ (the probability of drawing a card of a specified suit from the remaining 39 cards).

Various other populations can be represented by selecting certain subsets from the deck of 52 cards. I give below some simple sets that I have found useful. In each case the value of X is the face value of the card, F is the number of cards with a given value, and P is the probability associated with a given value, if a card is selected at random from the set. The Ace (A) is assigned a value of 1.

1. A uniform population:

Value	A	2	3	4	5
F	4	4	4	4	4
P	0.2	0.2	0.2	0.2	0.2

2. A skewed population:

Value	A	2	3	4
F	4	3	2	1
P	0.4	0.3	0.2	0.1

3. A U-shaped population:

Value	A	2	3	4
F	4	1	1	4
P	0.4	0.1	0.1	0.4

4. A symmetrical population:

Value	A	2	3	4	5
F	1	2	4	2	1
P	0.1	0.2	0.4	0.2	0.1

The above populations can be used to investigate the sampling distributions of means, sums, differences, and variances of samples drawn with and without replacement.

Table I in the Appendix is a table of 5000 random numbers in which the digits $0, 1, 2, 3, \ldots, 9$ appear with approximately the same frequency and in a random order. This table can also be used to investigate the sampling distributions of means, sums, differences, and variances of samples drawn from various populations. The populations described above, for example, can be represented by letting the random digits correspond to various values of X in the manner shown below.

1. A uniform population:

Random digit	0, 1	2, 3	4, 5	6, 7	8, 9
Value of X	1	2	3	4	5
P	0.2	0.2	0.2	0.2	0.2

2. A skewed population:

Random digit	0, 1, 2, 3	4, 5, 6	7, 8	9
Value of X	1	2	3	4
P	0.4	0.3	0.2	0.1

3. A U-shaped population:

Random digit	0, 1, 2, 3	4	5	6, 7, 8, 9
Value of X	1	2	3	4
P	0.4	0.1	0.1	0.4

4. A symmetrical population:

Random digit	0	1, 2	3, 4, 5, 6	7, 8	9
Value of X	1	2	3	4	5
P	0.1	0.2	0.4	0.2	0.1

For relatively small samples of size n drawn from the population of $N =$ 5000 random numbers, the sampling may, for all practical purposes, be assumed to be with replacement. It is also obvious that the table of random numbers can be used to represent various binomial populations.

In experiments with any of the sampling devices described above, prior probabilities can be assigned to each possible outcome. To most people, these probabilities seem reasonable under the assumptions stated. In some cases, however, there may be considerable disagreement as to the prior probabilities to be assigned to the possible outcomes of an experiment. Consider, for example, the probability of an ordinary thumbtack falling with its point up when dropped on a hard surface from a height of approximately 12 inches. The probabilities assigned to this event by students in my class have ranged from 0.05 to 0.95.

Suppose you guess that the prior probability of a thumbtack falling with its point up is 0.10. This guess may be regarded as a hypothesis. You decide to test your hypothesis by an experiment in which a thumbtack is dropped 100 times. Assuming your hypothesis to be true, it is possible to assign a probability to each possible outcome of the experiment. If the thumbtack falls with its point up 60 or more times, then it is easy to show that this is a highly improbable outcome under the hypothesis that the probability of a thumbtack falling with its point up is 0.10.

You may decide that the outcome is so improbable that your original hypothesis is incorrect. Or you may decide that the outcome is simply one of those improbable outcomes that will occur by chance, even when P is actually 0.10. Or your confidence in your original hypothesis may be sufficiently shaken that you decide to repeat the experiment an additional number of times to see whether you consistently obtain improbable outcomes. Whatever decision you make must inevitably depend on, among other things, how strongly you believe in your original hypothesis.

The above example describes, in essence, the manner in which outcomes of many experiments in the behavioral sciences are evaluated. A scientist proposes a hypothesis. Probabilities can then be assigned to the possible outcomes of an experiment, under the assumption that the hypothesis is true. If the particular outcome of the experiment is one that has a small probability of occurring, if the hypothesis is true, then the experimenter may decide to reject or modify his original hypothesis. Or he may decide to repeat the experiment an additional number of times. Whichever decision he makes will depend on, among other things, how strongly he believes in his original hypothesis.

It must be emphasized, however, that the outcome of any single experiment can never be conclusive. It may be convincing, but it cannot be conclusive. Improbable outcomes will occur, even when a hypothesis is correct. And it is also possible to obtain an outcome that is regarded as probable under a given hypothesis, even though the hypothesis is incorrect. A single improbable outcome of an experiment may convince a scientist that his original hypothesis was incorrect, but would it not be even more convincing if the experiment were repeated and if the outcome in this repetition were also improbable?

Judging from published research, it appears to be common practice to reject a hypothesis if the outcome of an experiment is such that it has a probability of 0.05 or less of occurring when the hypothesis is true. This practice is followed in subsequent chapters in testing hypotheses, not because I condone or approve of it, but simply because it would be awkward to repeat after each example the points made in the above paragraph. I hope, however, that in reading these chapters you will keep in mind the points I have emphasized above.

REVIEW OF TERMS

    binomial distribution
    binomial population
    binomial probability distribution

REVIEW OF SYMBOLS

N	σ_T	P	$\sqrt{nPQ}$
σ	σ_p	T	$\sqrt{PQ}$
μ_T	p	n	
σ^2/n	$\sigma_{\bar{X}}$	Q	

QUESTIONS AND PROBLEMS

9.1   We have a binomial population with $P = 1/3$ and $Q = 2/3$. We draw a random sample of $n = 10$ observations from this population. (a) In how many ways can we obtain a sample with $T = 8$? (b) What is the probability associated with any one of these samples in which $T = 8$? Do *not* multiply out the answer. (c) What is the probability of obtaining a sample with $T = 8$? Do *not* multiply out the answer. Explain how you obtain this probability.

9.2   Substitute in $(P+Q)^n$ with $P = 1/4$, $Q = 3/4$, and $n = 6$ and expand the binomial.

9.3   Let T be the number of values with $X = 1$ in Problem 9.2 (a) What is the value of μ_T? (b) What is the value of σ_T^2?

9.4   If random samples of $n = 18$ are drawn from the population of Problem 9.1 and if T is the number of values with $X = 1$, then what will be the value of μ_T and σ_T^2?

9.5   Assume that P is equal to 1/2 and that n is equal to 10. (a) What is the value of μ_T? (b) What is the value of σ_T^2? (c) What can you say about the shape or the form of the distribution of T?

9.6 We draw random samples of $n = 5$ observations from a binomial population in which $P = 1/3$. What can you say about the shape or form of the distribution of T?

9.7 We toss $n = 3$ coins and find T, the number of heads that are face up. Assume that the coins are unbiased so that for each coin the probability of a head is $1/2$. (a) Make a distribution showing the possible values of T and the probability associated with each value. (b) Multiply each value of T by its corresponding probability. The sum of the products will give the value of μ_T, and this should be equal to nP. (c) Calculate the variance of T from the distribution of T. This value should be equal to nPQ.

9.8 A fair coin is tossed $n = 10$ times. We assume that the probability of a head on each toss is $1/2$. Let T be the number of heads in a set of $n = 10$ tosses. (a) In how many ways can a value of $T = 10$ be obtained? (b) In how many ways can a value of $T = 9$ be obtained? (c) In how many ways can a value of $T = 8$ be obtained? (d) What is the probability of obtaining $T \geqslant 8$?

9.9 An unbiased die is rolled 36 times. We let T be the number of 6's that are obtained in a sample of $n = 36$ rolls. (a) What is the value of μ_T? (b) What is the value of σ_T^2? In answering the following questions, do *not* multiply out the terms involved in obtaining the correct answer. Simply give the terms so that if they were multiplied out the result would be the correct answer. (c) In $n = 36$ rolls, what is the probability of obtaining $T = 22$? (d) In $n = 36$ rolls, what is the probability of obtaining $T = 5$?

9.10 The following questions refer to drawing a random sample of $n = 4$ cards with replacement after each draw from a bridge deck of 52 cards. In answering the questions do *not* multiply out your answers. Simply give the numbers such that if they were multiplied out, the result would be the correct answer. Order does not count in any of the questions. (a) What is the probability that the sample will consist of four red cards? (b) What is the probability that the sample will consist of four spades? (c) What is the probability that the sample will consist of two spades and two cards of some other suit? (d) What is the probability that the sample will consist of two aces and two cards that are not aces? (e) What is the probability that the sample will consist of one king and three cards that are not kings?

9.11 In Problem 7.6 we described a discrimination experiment in which P was equal to $1/3$ and a subject was given $n = 5$ trials. We let T be the number of correct discriminations and you were asked to find the number of ways in which each possible value of T could occur. In Problem 7.7 you were asked to find the probability associated with each possible value of T. (a) Use these probabilities to find μ_T. (b) Explain why the value you obtain for μ_T should be equal to nP. (c) Now use the probabilities to find σ_T^2. (d) Explain why the value you obtain for σ_T^2 should be equal to nPQ.

9.12 In a discrimination experiment, the probability of a correct discrimination on a single trial is $P = 0.2$, under the assumption that a subject is responding by chance. A group of $n = 150$ independent subjects is tested. Let T be the number of subjects who make the correct discrimination. Find the value of μ_T and σ_T^2.

9.13 Write out the successive terms of the binomial expansion, $(P+Q)^n$, for the case of $P = 1/5$ and $n = 6$.

9.14 We have a finite binomial population with $N = 3$ elements in which one element has the value $X = 1$ and two elements have the value $X = 0$. (a) What are the values of μ and σ^2 for this population? (b) A random sample of $n = 2$ elements is drawn from this population *without* replacement. Let T be the sum of the values of X. Find the value of μ_T and σ_T^2.

9.15 We have a binomial population with $P(X = 1) = 3/4$. Random samples of $n = 3$ are drawn from this population. Find the probability distribution of T and use this distribution to calculate μ_T and σ_T^2. Show that these values are equal to nP and nPQ, respectively.

9.16 We have a finite binomial population with $N = 5$ elements of which three elements have the value of $X = 1$ and two have the value of $X = 0$. A random sample of $n = 2$ elements is drawn from this population without replacement. (a) Find the probability distribution of T and use this distribution to calculate μ_T and σ_T^2. (b) Show that μ_T is equal to nP and that σ_T^2 is equal to $[(N-n)/(N-1)]nPQ$.

10

The Normal Distribution

10.1 Introduction

One of the most useful theoretical distributions in statistical methods is the normal distribution. One reason for the importance of this distribution is the fact that, if we have a variable that is normally distributed in the population and if we draw random samples from this population, then the distribution of the sample sums T and the samples means $\bar{X}$ will also be normally distributed. This is so, as we shall see, even when a variable is *not* normally distributed in the population. If a variable is not normally distributed in the population, the distribution of the sample values of T and $\bar{X}$ for random samples drawn from the population will tend to be normally distributed as the sample size n is increased. Various other statistics, though not all, that are based on random samples also tend to be normally distributed as n increases. The range, for example, does not.

In the discussion and use of the normal distribution in this chapter, we shall assume that we are dealing with a variable that is normally distributed in the population with mean equal to μ and standard deviation equal to σ. It is to be emphasized, however, that the notion of normality applies to the shape or form of the distribution and not to any other properties that the distribution may have, such as the mean and standard deviation. We shall also assume that the distribution of interest is continuous, so that the shape or form may be represented by an equation for a curve.

10.2 The Standard Normal Curve

If two or more distributions are normal, they may still differ in terms of means and standard deviations. We can, however, transform any normally distributed variable into a new variable with mean equal to zero. For example, if we express each value of X as a deviation from the population mean, so that we have $x = X - \mu$, then we know that the distribution of x will have a mean of zero. Thus, if two or more variables are normally distributed with different means, we can transform the values of each variable to obtain new variables, all of which will have a mean equal to zero.

If the variables differ also in their standard deviations, then we can make an additional transformation that will result in new variables each with a standard deviation equal to 1.0. Thus, if we define

$$Z = \frac{X - \mu}{\sigma} \tag{10.1}$$

then it can be shown that

$$\mu_Z = 0$$

and

$$\sigma_Z = 1.0$$

Furthermore, if X is normally distributed, then Z will also be normally distributed. When this is the case, then Z is described as a *standard normal deviate* or, more briefly, a *normal deviate*.

If Z is normally distributed, then the distribution can be represented by a curve and the equation for this curve is

$$y = \frac{1}{\sqrt{2\pi}} e^{-(1/2)Z^2} \tag{10.2}$$

where $y = $ the height of the curve at any given point along the baseline
 $\pi = 3.1416$ (rounded), the ratio of the circumference of a circle to its diameter
 $e = 2.7183$ (rounded), the base of the natural system of logarithms
 $Z = (X - \mu)/\sigma$

The curve defined by (10.2) is called the *unit normal curve* or the *standard normal curve* and is illustrated in Figure 10.1. The area under the curve is equal to 1.00. Table III in the Appendix is a table of the unit normal curve. Column (1), headed Z, gives values of Z to two decimal places in units of 0.01. The second column gives the proportion of the total area between $\mu_Z = 0$ and a given value of Z. The third column gives the proportion of the total area in the *larger* segment of the curve, and the fourth column gives the proportion of the total area in the *smaller* segment. The fifth column gives the values of y corresponding to the values of Z, as obtained from (10.2). Because the normal curve is symmetrical about μ_Z equal to zero, the tabled values of Z are given only for positive values. The table can, however, be used with negative as well as with positive values of Z, because of the symmetrical distribution of Z.

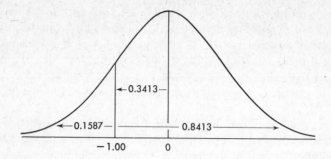

Figure **10.1** — A normal distribution showing the proportion of the total area falling below $Z = -1.00$ (0.1587), the proportion of the total area falling above $Z = -1.00$ (0.8413), and the proportion of the total area falling between the mean and $Z = -1.00$ (0.3413).

We illustrate the various relationships described above in Figure 10.1 with $Z = -1.00$. The proportions entered in the figure were obtained from Table III. Thus we see that 0.3413 of the total area falls between $Z = -1.00$ and the mean μ_Z, which is equal to zero. We see that 0.8413 of the total area falls above $Z = -1.00$, that is, in the larger segment of the curve, and that 0.1587 of the total area falls below $Z = -1.00$, that is, in the smaller segment.

It is important to note that areas of the normal curve can be interpreted as probabilities. For example, if Z is a normally distributed variable, and if a value of Z is selected at random, then the probability of obtaining $Z \leqslant -1.00$ is 0.1587 and the probability of obtaining $Z \geqslant -1.00$ is 0.8413. Similarly, the probability of obtaining a value of Z equal to or less than zero but equal to or greater than -1.00 is 0.3413.

10.3 Applications of the Standard Normal Curve

Assume that for a population of ten-year old children, IQ's are normally distributed with $\mu = 100$ and $\sigma = 15$. What proportion of these children will have IQ's within the inclusive range 90 to 110? In other words, using the standard normal curve we wish to find $P(90 \leqslant \text{IQ} \leqslant 110)$. Because the IQ's are discrete we use the lower limit 89.5 of 90 and the upper limit 110.5 of 110.[1] Transforming these values into standard normal deviates, we have

$$Z = \frac{89.5 - 100}{15} = -0.70 \quad \text{and} \quad Z = \frac{110.5 - 100}{15} = 0.70$$

From the table of the standard normal curve we find that 0.2580 of the total area falls between the mean μ and $Z = 0.70$. Because the curve is

---

[1]The use of a lower (or upper) limit for a discrete value rather than the discrete value itself, is called a *correction for discreteness* or a *correction for discontinuity*.

symmetrical, we also have 0.2580 of the total area falling between the mean μ and $Z = -0.70$. Then $0.2580 + 0.2580 = 0.5160$ or approximately 52 percent of the children will have IQ's within the inclusive range 90 to 110.

If an IQ of 70 or lower is regarded as indicating "feeblemindedness," what proportion of the children would be described as feebleminded? Because in this case we want to find $P(IQ \leq 70)$, we use the upper limit 70.5 of the discrete value of 70. Then we have

$$Z = \frac{70.5 - 100}{15} = -1.97$$

and we find from the table of the standard normal curve that 0.0244 of the total area falls to the left of $Z = -1.97$ and this is the proportion of the children with IQ's of 70 or lower.

Suppose that a random sample of $n = 100$ children is selected from the population. What is the probability that the sample mean $\bar{X}$ will fall within the inclusive range 97.00 to 103.00? The standard error of the mean of a random sample of $n = 100$ will be

$$\sigma_{\bar{X}} = \frac{\sigma}{\sqrt{n}} = \frac{15}{\sqrt{100}} = 1.5$$

If the IQ's are normally distributed, then the sample means will also be normally distributed about the population mean $\mu = 100$. Then

$$Z = \frac{\bar{X} - \mu}{\sigma_{\bar{X}}}$$

will be a standard normal deviate and can be evaluated in terms of the standard normal curve.

In this case, we wish to find, using the standard normal curve, $P(97.00 \leq \bar{X} \leq 103.00)$. The sample means can differ from one another in units of 0.01, and the lower limit of 97.00 is 96.995 and the upper limit of 103.00 is 103.005. Then

$$Z = \frac{96.995 - 100}{1.5} = -2.00 \quad \text{and} \quad Z = \frac{103.005 - 100}{1.5} = 2.00$$

rounded to two decimal places. From the table of the standard normal curve, we find that $0.4772 + 0.4772 = 0.9544$ of the total area falls within the range $Z = -2.00$ and $Z = 2.00$. Thus if we have a random sample of $n = 100$ children, the probability that the sample mean will fall within the inclusive range 97.00 to 103.00 is 0.9544.

10.4 Use of the Standard Normal Curve To Evaluate the Outcome of a Sample Drawn from a Binomial Population

Now let us see how we can apply the standard normal distribution in evaluating the results of a binomial experiment we have conducted.

Suppose, for example, we lunch with a friend who claims that he can discriminate his favorite brand of cigarettes, which we will call "A," from another brand, which we will call "B." We decide to test his claim by a series of observations, and he agrees to participate in the research. Each day, after lunch, we blindfold the friend and give him two cigarettes, one of which is Brand "A" and the other Brand "B." After he has smoked both cigarettes, he informs us which one he believes was Brand "A." This verbal report is the observation and we assign the value of $X = 1$ to the observation if he has made a correct choice, and a value of $X = 0$ if he has not. We carry the experiment on over a series of days, making one observation each day, until we have made $n = 100$ observations. We let T be the number of correct discriminations. Assume that the outcome of the experiment is as given in Table 10.1. We have $T = \Sigma X = 69$ and $\bar{X} = T/n = P = 0.69$.

Table **10.1** — Distribution of Correct and Incorrect Choices in a Discrimination Experiment

	T	p
Correct choices	69	0.69
Incorrect choices	31	0.31
Total	100	

We now consider a theoretical model that can be used in evaluating the outcome of the experiment. In this theoretical model, we shall assume that the friend is *not* able to discriminate between the two brands of cigarettes. If he is unable to do so, then we assume that his verbal report is no better than a guess. If the friend is just guessing, he should guess correctly no more often then he guesses wrong. Consider a population of such guesses. For this population, we would have $P = 0.5$ as the probability of a correct guess and $Q = 1 - P = 0.5$ as the probability of a wrong guess.

Note that we are dealing with a *theoretical* binomial population. We know that for this theoretical binomial population we have $\sigma^2 = PQ = (0.5)(0.5)$. Now let us assume that we draw random samples of $n = 100$ observations from this binomial population and we count the number of values of $X = 1$ in this sample. We know that the long-run average value of $T = \Sigma X$ will be given by

$$\mu_T = nP$$

where P is the probability of obtaining a value of $X = 1$. We also know that the variance of T will be given by

$$\sigma_T^2 = n\sigma^2$$

where $\sigma^2 = PQ$ or the variance of the binomial population. Then, for this theoretical distribution, we have

$$\mu_T = (100)(0.5) = 50$$

and

$$\sigma_T{}^2 = (100)(0.5)(0.5) = 25$$

so that

$$\sigma_T = \sqrt{25} = 5$$

The possible outcomes of the experiment consist of the possible values that T can take and these are $0, 1, 2, \ldots, 100$. We might consider expanding

$$(P+Q)^n = (0.5+0.5)^{100}$$

to determine the probability of obtaining $T \geq 69$, but this would obviously be exceedingly tedious.

If you will go back to the previous chapter and look at the distribution of T when $P = 0.5$ and when $n = 10$, as shown in Figure 9.2, you will observe that T appears to have a distribution that might be approximated by a normal distribution. Let us assume that this is, in fact, the case in the present experiment. Thus, if T is approximately normally distributed, then

$$Z = \frac{T - \mu_T}{\sigma_T} \tag{10.3}$$

will be a standard normal deviate and can be evaluated in terms of the table of the standard normal curve. Note the similarity between Z as defined above and Z as given by (10.1). In the one case, we assume that X is normally distributed and in the other that T is normally distributed. We also have μ as the mean of the distribution of X and μ_T as the mean of the distribution of T, and we have σ as the standard deviation of the X distribution and σ_T as the standard deviation of the T distribution.

In the present example, we have $T = 69$, $\mu_T = 50$, and $\sigma_T = 5$. Because $T = 69$ is discrete and we want to find the probability of $T \geq 69$, we use the lower limit, 68.5, of the discrete value. Then

$$Z = \frac{68.5 - 50}{5} = 3.70$$

and from the table of the standard normal curve we find that $P(Z \geq 3.70) = 0.0001$ and this is an estimate of the probability of obtaining T equal to or greater than 69 in a random sample of $n = 100$ observations from a binomial population in which $P = 0.5$. Thus, under the assumptions we have made, if the friend is just guessing as to which of the two cigarettes is Brand "A," the outcome of the experiment is one that could be expected to occur only one time in 10,000 as a result of random sampling. We might conclude, therefore, that either an extremely improbable event has occurred or our theoretical model is in error. It is not difficult to show that the values of T, based on random samples of $n = 100$ observations drawn at random from a binomial population in which $P = 0.5$, are distributed in a manner very closely approximating a normal distribution. The error is not in this assumption, nor is it with the value of σ_T, because this is fixed once P and n are known. The source of the difficulty lies in the value we assumed for P.

If we retain all of the assumptions we have made, but assume that P is equal to 0.65, then we would have $\sigma_T = \sqrt{(100)\,(0.65)\,(0.35)} = 4.8$. Then

$$Z = \frac{68.5 - 65}{4.8} = 0.73$$

and, from the table of the standard normal curve, we find that the probability of Z equal to or greater than 0.73 is 0.2327. Thus, in random sampling from a binomial population in which $P = 0.65$, a value of T equal to or greater than 69 would not be too unusual in a sample of $n = 100$ observations. Values of T equal to or greater than 69 would be expected to occur about 20 times in 100.

But if we assume P to be greater than 0.5, we are also assuming that the friend is able to make some degree of discrimination between the two brands of cigarettes; that is, he will be right more often than he is wrong. If we had granted this point to begin with, there would have been no need to make the observations or to conduct the experiment. Thus it would intuitively seem that the most reasonable value to take for P, in evaluating the outcome of the experiment, is $P = 0.5$. The fact that the outcome of the experiment results in T equal to 69 and that values as large as or larger than this would occur only one time in 10,000 by chance, if P is actually equal to 0.5, leads us to doubt the initial assumption of $P = 0.5$.

In making use of (10.3) for the problem described, we assumed that the distribution of T for random samples of $n = 100$ observations drawn from a binomial population with $P = 0.5$ was approximately normal in form. In general, this assumption is tenable if both nP and nQ are equal to or greater than 5.

10.5 The Evaluation of p

If we divide both the numerator and denominator of (10.3) by n, the number of observations, we have

$$Z = \frac{\dfrac{1}{n}\,(T - \mu_T)}{\dfrac{1}{n}\,\sigma_T}$$

But $T/n = p$, $\mu_T = nP$ and, therefore, $\mu_T/n = P$. Also $\sigma_T = \sqrt{nPQ}$ and, therefore, $\sigma_T/n = \sqrt{PQ/n}$. Thus, we also have

$$Z = \frac{p - P}{\sqrt{\dfrac{PQ}{n}}} \tag{10.4}$$

For the experiment under discussion, we wanted to find $P(T \geqslant 69)$ in terms of a normal distribution and we used the lower limit of T or 68.5. Similarly, we use $p = 68.5/100 = 0.685$ in (10.4) rather than $p = 69/100 = 0.69$. Then we have

$$Z = \frac{0.685 - 0.500}{\sqrt{\dfrac{(0.5)\,(0.5)}{100}}} = \frac{0.185}{0.05} = 3.70$$

as before.

It does not matter, therefore, whether we evaluate T or p as the outcome of the experiment. Regardless of which we choose to evaluate, we will obtain exactly the same value of Z.

10.6 The Central-Limit Theorem

We have seen that the standard normal curve can be used to evaluate either the sum T or the mean p of a random sample drawn from a binomial population. Note that a binomial population in which X can take only the values of 1 or 0 is *not* a normally distributed population. But the values of T or p, based on random samples from a binomial population, do tend to have a distribution that is reasonably normal in form. The fact is that, regardless of the shape or form of a population distribution, the distributions of both the sum and the mean of random samples approach that of a normal distribution as the sample size is increased. This statement is based on an important theorem known as the *central-limit theorem*.

Suppose, for example, that a variable X can take only the values of 0, 1, 2, or 3, with corresponding probabilities of 0.4, 0.1, 0.1, and 0.4. The population distribution of X is U-shaped. Yet, it can be shown that if random samples of $n = 8$ are drawn from this population, the distribution of T and $\bar{X}$ will be reasonably normal in form. As n is increased, the distributions of both T and $\bar{X}$ tend to approach more closely a normal distribution. This is obviously an extreme case, simply because relatively few, if any, population distributions of measurements in psychology or education are U-shaped. At worst, the variable of interest may have a skewed or rectangular distribution. For skewed and rectangular distributions, the distribution of the sample values of T and $\bar{X}$ approaches that of a normal distribution with relatively small values of n.

The import of the central-limit theorem is that we need not be overly concerned about the statement that X is assumed to be normally distributed. We shall not be primarily concerned with the distribution of X, but rather with the distribution of T or $\bar{X}$. If we have an adequate number of observations, then it will be reasonable to assume that both T and $\bar{X}$ are approximately normally distributed, regardless of the nature of the distribution of X.

REVIEW OF TERMS

central-limit theorem   standard normal curve
correction for discontinuity   standard normal distribution
normal deviate

REVIEW OF SYMBOLS

Z	P	μ_Z
σ_T	p	σ
μ_T	σ_p	σ_Z

QUESTIONS AND PROBLEMS

10.1   What is the difference between Z defined as a standard normal deviate and z defined as a standard score?

10.2   Assume that X is a variable corresponding to the number of correct responses on a test and that the values of X are approximately normally distributed with $\mu = 50.0$ and $\sigma = 10.0$. In using the table of the normal curve to find $P(X \geq 40)$ we would use the lower limit 39.5 because we want the area in the right tail to *include* the value of 40 for which the lower limit is 39.5. To find $P(X > 40)$, we would use the upper limit 40.5 because we want the area in the right tail to *exclude* 40. To find $P(X \leq 40)$ we would use the upper limit 40.5 because we want the area in the left tail to *include* 40. To find $P(X < 40)$ we would use the lower limit 39.5 because we want this area to *exclude* 40. For each of the following, substitute the appropriate upper or lower limits that should be used.

(a) $P(X > 60)$ (e) $P(25 \leq X \leq 75)$
(b) $P(X \geq 60)$ (f) $P(25 < X \leq 75)$
(c) $P(X < 25)$ (g) $P(25 \leq X < 75)$
(d) $P(X \leq 30)$ (h) $P(25 < X < 75)$

10.3   We draw a single observation at random from the population of Problem 10.2. We are interested in the value of X such that:
(a) The probability is 0.0495 of drawing $X \geq$?
(b) The probability is 0.0322 of drawing $X \geq$?

10.4   We have a binomial population with $P = 0.5$ and $Q = 0.5$. We have a random sample from this population of $n = 20$ observations. What is the probability that T will be equal to or greater than 15? Determine the probability by assuming that T is approximately normally distributed. Calculate the probability both with and without a correction for discontinuity.

10.5   In a discrimination experiment, a subject is asked to state which of three exposed lines is the longest. Assume that his response is a matter of chance and therefore that the probability of a correct discrimination is 1/3. The subject is given n trials, and we assume that the probability of a correct discrimination remains the same on each trial. (a) If we are to make use of the table of the standard normal curve in evaluating the outcome of the experiment, what is the minimum number of trials that the subject should be given? (b) Suppose that the subject is given 30 trials and we find that he makes $T = 16$ correct discriminations. What is the probability of 16 or more correct discriminations under the conditions stated? Determine the probability using the table of the standard normal curve. (c) If we make a correction for discontinuity, we would use $T = 15.5$ rather than 16. Explain why this is the proper correction for the problem described. (d) Determine the probability using a correction for discontinuity.

10.6   In Problem 8.3 it was possible for you to determine the exact probability of a sample mean equal to or greater than 4.5. (a) What is this exact probability? (b) What is the standard error of the mean for this same problem? (c) Assume that the sample means are approximately normally distributed. Find $Z = (4.5 - \mu)/\sigma_{\bar{x}}$ and determine the probability associated with Z equal to or larger than the value obtained. (d) We know that the values of the sample means are discrete. For example, we can have a mean of 4.5 or of 4.0, but no mean can fall between these two values. Explain why, if we make a correction for discontinuity in determining the probability of a sample mean equal to or greater than 4.5, we should use 4.25 as the corrected value. (e) Find the value of Z corresponding to $(4.25 - \mu)/\sigma_{\bar{x}}$ and determine the probability associated with Z equal to or greater than the value obtained. (f) Is the probability obtained using a correction for discontinuity more in accord with the known exact probability than that determined without using a correction for discontinuity?

10.7   In a random sample of 100 college students, 63 answered "Yes" to the question: "Would you make use of the library facilities if the building were open on Sunday afternoon?" Use the table of the standard normal curve to determine whether it is reasonable to believe that this sample was drawn from a binomial population in which the probability of a "Yes" response is 0.5. Make a correction for discontinuity.

10.8   In a random sample of $n = 50$ students, 35 of them said they were going to vote for Candidate "A" in a student election. Does it seem reasonable to believe that in the population of students the probability of a vote for Candidate "A" is equal to 0.5? Make a correction for discontinuity. Explain your answer.

10.9   A die is rolled 36 times. We let T be the number of 6's obtained in $n = 36$ rolls. (a) What is the value of μ_T? (b) What is the value of σ_T^2? (c) Use the table of the standard normal distribution to determine the probability of $T \geq 9$. Make a correction for discontinuity.

10.10 A fair die is rolled $n = 180$ times. Let T be the sum of the numbers face up. (a) Calculate μ_T and σ_T. (b) Assume that T is normally distributed. Use the table of the standard normal curve to find the probability of $T \leq 600$. Make a correction for discontinuity.

10.11 In a discrimination experiment the probability of a correct response on a single trial is $P = 2/5$, under the assumption that a subject is responding by chance. A group of $n = 150$ subjects is tested. Let T be the number of subjects who make the correct response. (a) Calculate μ_T and σ_T. (b) Assume that T is normally distributed. Use the table of the standard normal curve to find the probability of $T \leq 44$. Make a correction for discontinuity.

10.12 Assume that X can take values of $1, 2, 3, \ldots, 25$, each with a probability of $1/25$. (a) Calculate μ and σ^2 for this population. (b) A random sample of $n = 5$ is drawn from the population *without* replacement. Let T be the sum of the n values of X. Calculate μ_T and σ_T^2. (c) Assume that T is normally distributed. Use the table of the standard normal curve to find the probability of $T \geq 75$. Make a correction for discontinuity.

10.13 On a multiple choice test the probability of making a correct response by chance is $P = 1/4$ for each of $n = 48$ items. Let T be the number of correct responses. (a) Under the assumption that responses to the items are a matter of chance, calculate μ_T and σ_T. (b) Find the probability of $T \geq 17$ correct responses using the table of the standard normal curve. Make a correction for discontinuity.

10.14 Assume that in a discrimination experiment the probability of a correct response on each trial is $P = 1/3$. A subject is given $n = 18$ independent trials. (a) Calculate μ_T and σ_T. (b) Use the table of the standard normal curve to find the probability of $T \geq 12$. Make a correction for discontinuity.

11

The t Test for Means

11.1 Introduction

In research and experimental work, we deal with a sample of n observations. We may consider the sample set of observations to be randomly and independently drawn from some population but, in general, the variance of this population is unknown. We do know, however, that the sample variance, which we have defined as

$$s^2 = \frac{\Sigma(X-\bar{X})^2}{n-1}$$

will be an unbiased estimate of the population variance σ^2, and that $\bar{X}$ is an unbiased estimate of μ, the population mean.

We need now to introduce another concept that has not been necessary in our discussion so far. The concept is that of *degrees of freedom*, which we will represent by d.f. The number of degrees of freedom associated with the variance, as defined above, is given by the denominator, $n-1$. In calculating the sum of squares, we take deviations, $X-\bar{X}$, from the sample mean. We have n of these deviations, but only $n-1$ of them are free to vary, because we know that $\Sigma(X-\bar{X})$ must equal zero. Thus, given the values of any set of $n-1$ of the n deviations, the value of the remaining one is completely determined. We say, therefore, that the sum of squares, $\Sigma(X-\bar{X})^2$, has $n-1$ d.f. This sum of squares, when divided by its degrees of freedom, provides an estimate of the population variance with $n-1$ d.f.

Using the estimate of the population variance, the variance of the means of random samples of size n will be given by

$$s_{\bar{X}}^2 = s^2/n$$

and the standard error of the mean will be

$$s_{\bar{X}} = s/\sqrt{n}$$

11.2 The *t* Distribution

If we let μ be the population mean and $\bar{X}$ the mean of a sample, then

$$t = \frac{\bar{X} - \mu}{s_{\bar{X}}} \tag{11.1}$$

has a known distribution that is called the *t* distribution. We note that *t*, as defined above, differs from what we have called a normal deviate *Z* in that we have used $s_{\bar{X}}$ in the denominator rather than $\sigma_{\bar{X}}$. When we use $s_{\bar{X}}$ as a standard error it, in turn, is based on an estimate s^2 of the population variance, whereas $\sigma_{\bar{X}}$ is based on a known value σ^2 of the population variance. Because the estimate of the population variance s^2 has $n-1$ d.f., the distribution of *t* will be different for different degrees of freedom.[1]

Table IV in the Appendix gives the probabilities with which various values of *t* will be equaled or exceeded in random sampling from a normally distributed population. The row of the table is entered with the number of degrees of freedom. The column headings give the probability of obtaining a *t* equal to or greater than the values tabled in the various rows for the number of degrees of freedom available. The distribution of *t* is symmetrical and the table can be used to determine the probability for either plus or minus values of *t* or for an absolute value. Take, by way of illustration, the distribution of *t* for 16 d.f. The mean of the distribution, regardless of the number of degrees of freedom, is zero. We move out in a positive direction from the mean until we come to *t* equal to 2.120. If we erect an ordinate at this point, we will cut off 0.025 of the total area in the right tail of the distribution. The column heading for *t* equal to 2.120 with 16 d.f. is 0.025. If we move out from the mean in a negative direction until we come to *t* equal to -2.120 and erect an ordinate at this point, we will cut off 0.025 of the total area in the left tail. The probability of an absolute value of *t* equal to or greater than 2.120 will be given by the sum of the areas in the two tails or $0.025 + 0.025 = 0.05$. Therefore, if we wish to determine the probability of an absolute value of *t* equal to or greater than the tabled value, we double the probability given by the column headings.

11.3 Testing a Hypothesis about μ

Let us assume that we have a set of $n = 17$ observations and that the mean of this sample is 7.12. Let us suppose also that the standard error of

---

[1] With an infinite number of degrees of freedom, s^2 will be equal to σ^2 and, in this case, the distribution of *t* is the same as the distribution of the normal deviate *Z*.

the mean, as given by $s/\sqrt{n}$, is 1.00. We do not know the value of the population mean, but we are interested in the hypothesis that it is 5.00. *If* it is 5.00, then how frequently would we obtain in random sampling a mean of 7.12 or greater? Substituting in (11.1), we have

$$t = \frac{7.12 - 5.00}{1.00} = 2.12$$

with 16 d.f. By reference to the table of t, we find that the probability of obtaining a t equal to or greater than 2.12 is 0.025. *If* the population mean is 5.00, then we should get sample means of 7.12 or greater only 25 times in 1000.

11.4 Confidence Limits for the Mean

Suppose that we have a set of $n = 17$ observations, and that the mean of this sample is 10.00. Let us also suppose that the standard error of the mean, as obtained by $s/\sqrt{n}$, is 2.00. We are interested in the unknown value of the mean in the normally distributed population from which the sample is assumed to have been randomly drawn. Our best estimate of the population mean is, of course, the sample mean, because we know that it is an unbiased estimate. But can we not say something more about the population mean than this? Intuitively, it would appear much more reasonable to believe that the population mean is, say 12.00 rather than 25.00. How can we express this intuitive belief more objectively?

The procedure commonly used is to specify an interval in which one has a certain degree of confidence that the parameter falls. This interval is called a *confidence interval* and the limits of the interval are called *confidence limits*. The degree of confidence that we have in the statement that a parameter falls within the confidence interval is called a *confidence coefficient*. Suppose we wish to determine a 95 percent confidence interval for μ in the problem described above. From the table of t, we find that for 16 d.f., values of t equal to or less than -2.12 will occur with a probability of 0.025. We designate this t as $t_{0.025}$. Because the t distribution is symmetrical, values of t equal to or greater than 2.12 will also occur with a probability of 0.025. If 0.025 is the probability of t equal to or greater than 2.12, we can also say that $1.00 - 0.025 = 0.975$ is the probability of t *less* than 2.12. We designate this t as $t_{0.975}$.

If we denote the lower confidence limit by c_1 and the upper confidence limit by c_2, we can solve for these two limits in terms of the following equations:

$$\frac{\bar{X} - c_1}{s_{\bar{X}}} = t_{0.975} \tag{11.2}$$

and

$$\frac{\bar{X} - c_2}{s_{\bar{X}}} = t_{0.025} \tag{11.3}$$

Solving for c_1 and c_2, with $\bar{X}$ equal to 10.00, $s_{\bar{X}}$ equal to 2.00, $t_{0.975}$ equal to 2.12, and $t_{0.025}$ equal to -2.12, we have

$$c_1 = 10.00 - (2.12)(2.00) = 10.00 - 4.24 = 5.76$$

and

$$c_2 = 10.00 + (2.12)(2.00) = 10.00 + 4.24 = 14.24$$

The values of 5.76 and 14.24 are the 95 percent confidence limits for the sample under consideration and define a 95 percent confidence interval for the sample. Because the confidence limits are statistics, they are subject to sampling variation like all other statistics. If we drew another sample of $n = 17$ observations from the same population as the first sample, both the mean and the standard error of the mean for this second sample might differ from the values we obtained with the first sample. Thus the confidence limits established by the second sample would not necessarily be the same as those established by the first sample. When we say that we are 95 percent confident that the population mean falls within the 95 percent confidence interval, we are expressing our confidence that, in repeated sampling, the inference concerning the population mean will be correct 95 times in 100; that is, on the average, 95 out of 100 confidence intervals established in the manner described will include the population mean. For any particular sample, this inference will either be right or wrong; that is, either the population mean falls within the interval or it does not.

11.5 Use of a Table of Random Numbers

Suppose that we have an experiment in which we are to test $n = 10$ subjects under Treatment 1 and $n = 10$ subjects under Treatment 2. It is essential, if we are to use the techniques to be described in evaluating the outcomes of this experiment, that the subjects be randomly assigned to the two experimental treatments. How can we do this?

Suppose that 500 students are registered in the introductory psychology course at a given university. A random sample of $n = 20$ is to be selected from the population of 500 subjects. We could select the sample by the method described earlier by writing the names of each of the 500 subjects on a disk. These disks could then be placed in a box, the disks thoroughly mixed, and a set of 20 disks could be drawn. This procedure, we believe, would provide a random sample, but, even so, it would be laborious. A simpler procedure would involve assigning one of the numbers 000, 001, 002, 003, . . . , 499 to each of the 500 subjects. It does not matter which subject receives which number; it is only necessary that different numbers identify different subjects. The selection of the sample is then made by means of a table of random numbers.

Table 1 in the Appendix is a table of random numbers. To find a point of entry into the table, one should use some random procedure. The table consists of five blocks of 1000 random numbers each. For each block, the rows have been numbered from 00 to 24 and the columns from 00 to 39. Thus, if we select a number from 1 to 5 at random, another from 00 to 24 at random, and a third from 00 to 39 at random, this will give a random point of entry into the table. Let us suppose that the point of entry into the table is 2, 02, and 05. Once we have entered the table, it makes no difference whether we read up, down, or across the table. Because the numbers assigned to the 500 subjects consist of three-digit numbers, we shall make use of columns 05, 06 and 07, and read down. We read down these columns, at the point of entry, selecting the first 20 unlike numbers between 000 and 499. We skip any number that exceeds 499 and any repetition of a number previously read from the table. The subjects assigned these 20 numbers will constitute a random sample of $n = 20$. By a flip of a coin we can then assign each subject to one of the two treatments, with the restriction that we have an equal number of subjects for each treatment.

11.6 Difference between Two Means

If we have a single random sample of n_1 observations drawn from a population with variance σ_1^2, the population variance will be estimated by s_1^2, and the variance of the mean will be estimated by s_1^2/n_1. If we have a second random sample of n_2 observations from a population with variance σ_2^2, the population variance will be estimated by s_2^2, and the variance of the mean will be estimated by s_2^2/n_2. The variance of the difference between the two means will be estimated by

$$ s_{\bar{X}_1 - \bar{X}_2}^2 = \frac{s_1^2}{n_1} + \frac{s_2^2}{n_2} \tag{11.4} $$

and the standard error of the difference between the means will be estimated by the square root of (11.4) or

$$ s_{\bar{X}_1 - \bar{X}_2} = \sqrt{\frac{s_1^2}{n_1} + \frac{s_2^2}{n_2}} \tag{11.5} $$

If the samples are from populations with the same variance, we may pool the sums of squares and degrees of freedom for the two samples to obtain an estimate of the common population variance. Thus

$$ s^2 = \frac{\Sigma(X_1 - \bar{X}_1)^2 + \Sigma(X_2 - \bar{X}_2)^2}{n_1 + n_2 - 2} \tag{11.6} $$

where $s^2 =$ the estimate of the common population variance
$\Sigma(X_1 - \bar{X}_1)^2 =$ the sum of squared deviations of the n_1 observations from the mean for Treatment 1

$\Sigma(X_2 - \bar{X}_2)^2$ = the sum of squared deviations of the n_2 observations from the mean for Treatment 2

The degrees of freedom for the estimate of s^2 are clearly indicated by the denominator, $n_1 + n_2 - 2$.

Substituting in (11.5) with the common estimate s^2, we obtain

$$s_{\bar{X}_1 - \bar{X}_2} = \sqrt{s^2\left(\frac{1}{n_1} + \frac{1}{n_2}\right)} \qquad (11.7)$$

and if $n_1 = n_2 = n$, then (11.7) may be written

$$s_{\bar{X}_1 - \bar{X}_2} = \sqrt{2s^2/n} \qquad (11.8)$$

Table 11.1 — Values of a Variable X Obtained in Two Samples of $n = 10$ Observations Each

	(1) Group 1	(2) Group 1	(3) Group 2	(4) Group 2
	X_1	X_1^2	X_2	X_2^2
	6	36	1	1
	3	9	3	9
	4	16	4	16
	5	25	2	4
	2	4	5	25
	5	25	5	25
	4	16	2	4
	6	36	4	16
	3	9	3	9
	2	4	1	1
Σ	40	180	30	110

Table 11.1 gives the values of X and X^2, the dependent variable, for the subjects in each of the two treatments. We see that

$$\bar{X}_1 = 40/10 = 4.0 \text{ and that } \bar{X}_2 = 30/10 = 3.0$$

The sum of squared deviations for Treatment 1 is

$$\Sigma(X_1 - \bar{X}_1)^2 = 180 - (40)^2/10 = 20$$

and, similarly, the sum of squared deviations for Treatment 2 is

$$\Sigma(X_2 - \bar{X}_2)^2 = 110 - (30)^2/10 = 20$$

Then the standard error of the difference between the two means, as given by (11.8), will be

$$s_{\bar{X}_1 - \bar{X}_2} = \sqrt{\frac{2}{10}\left(\frac{20 + 20}{10 + 10 + 2}\right)} = 0.67$$

We now define t as

$$t = \frac{\bar{X}_1 - \bar{X}_2 - (\mu_1 - \mu_2)}{s_{\bar{X}_1 - \bar{X}_2}}$$ (11.9)

where t = the t ratio with $n_1 + n_2 - 2$ d.f.

$\bar{X}_1$ = the mean for Treatment 1

$\bar{X}_2$ = the mean for Treatment 2

μ_1 = the population mean of which $\bar{X}_1$ is an estimate

μ_2 = the population mean of which $\bar{X}_2$ is an estimate

We now state, as a hypothesis to be tested, that the observations in the two samples have been independently and randomly drawn from identical normally distributed populations. If this hypothesis is true, then $\sigma_1^2 = \sigma_2^2$ and $\mu_1 = \mu_2$. Therefore, $\mu_1 - \mu_2$ will be equal to zero. Setting $\mu_1 - \mu_2$ equal to zero and substituting in (11.9) we have, for the data of Table 11.1,

$$t = \frac{4.0 - 3.0}{0.67} = 1.49$$

with $n_1 + n_2 - 2 = 18$ d.f.

The value of t we have just obtained provides us with a basis for evaluating the hypothesis stated for, if the hypothesis is true, the expected or average value of t should be zero. By entering the table of t with 18 d.f., we find that values of t equal to or greater than 1.49 would occur, when $\mu_1 = \mu_2$, with a probability of less than 0.10 but greater than 0.05. Similarly, the probability of obtaining values of t equal to or less than -1.49 is also less than 0.10 but greater than 0.05, when $\mu_1 = \mu_2$. Consequently, if the null hypothesis is true, the probability of obtaining an *absolute* value of t equal to or greater than 1.49 is less than 0.20 but greater than 0.10.

11.7 Tests of Significance

The example we have just described contains all of the essential elements of a *test of significance*. We have also used tests of significance previously, but they have not been described as such. A test of significance provides us with a basis for evaluating some stated hypothesis. In essence, the test of significance yields a probability. If the probability associated with the test is small, then we conclude that *either* an exceptionally rare event has occurred, *or* that the hypothesis we have tested is false.

The hypothesis that we test is called a *null hypothesis*. In the example described, the null hypothesis was that the two sets of observations were randomly and independently drawn from identical normally distributed populations. We proceeded by regarding the null hypothesis as true. If it is true, then the theoretical distribution of t, as obtained from

(11.9), is known exactly and specifies completely the probabilities associated with the possible outcomes of the experiment. Thus, we use the *t* distribution to determine the probability of obtaining an absolute value of $\bar{X}_1 - \bar{X}_2$ equal to or greater than the observed value under the null hypothesis that $\mu_1 = \mu_2$. If this probability is small, we conclude that either a very rare event has occurred or that the hypothesis $\mu_1 = \mu_2$ is false. Our interest, in other words, is in whether or not the observations in the two samples offer evidence against or contradict the null hypothesis. The null hypothesis, in this sense, is one which the observations may or may not void or nullify.

11.7.1 LEVELS OF SIGNIFICANCE

In testing hypotheses, we must decide what we shall choose to call a small probability. There are many technical and theoretical facets to this question, and the choice is best determined by considering the particular problem under investigation. We shall, however, need some guidepost in subsequent discussions and, for this purpose only, we shall choose a value that is frequently used by research workers. We shall regard a small probability as one that is equal to or less than 0.05. If, in evaluating a statistic, we find that the test of significance results in a probability of 0.05 or less, assuming the null hypothesis to be true, then the outcome we have obtained would occur five times or less in 100, as a result of chance or random sampling. When this occurs, we say that we reject the null hypothesis tested. The probability that we have chosen to use in rejecting null hypotheses is called the *level of significance* of the test and is symbolized by α. If we choose α equal to 0.05 and the test of significance results in a probability of 0.05 or less, we say that the result is significant at the 5 percent level.

For the data of Table 11.1 we have $n_1 + n_2 - 2 = 10 + 10 - 2 = 18$ d.f. for *t*. From the table of *t* we find that with 18 d.f. the probability of obtaining *t* equal to or greater than 2.101 is 0.025 and similarly the probability of obtaining *t* equal to or less than -2.101 is 0.025, when $\mu_1 = \mu_2$. The probability of obtaining an *absolute* value of *t* equal to or greater than 2.101 is therefore $0.025 + 0.025 = 0.05$. If the significance level of our test is $\alpha = 0.05$, then either an obtained $t \geq 2.101$ or an obtained $t \leq -2.101$ would result in rejection of the null hypothesis $\mu_1 = \mu_2$. Because the value of *t* we obtained was only equal to 1.49, it was not significant at the 5 percent level. In other words, for this experiment the null hypothesis $\mu_1 = \mu_2$ would not be rejected with a significance level $\alpha = 0.05$.

We emphasize that many other values of α rather than 0.05 could be used as the significance level. If we make α very small, however, we must realize that we decrease the probability of rejecting the null hypothesis. For example, if we choose α equal to 0.000001, we would be demanding that the outcome of the experiment be one that could be expected to occur only one time in 1,000,000, when the null hypothesis is true. If we make α sufficiently small, then no outcome of the experiment would ever be considered as offering significant evidence against the null hypothesis, even if it were, in fact, false.

11.7.2 Two Types of Errors

When a null hypothesis is true and the results of a test of significance reject it, or declare it false, we describe this as a *Type I error*. The probability of a Type I error is equal to α, the significance level we have chosen. By making α small, we have good protection against making a Type I error, but, at the same time, we increase the probability of making a Type II error. A *Type II error* occurs when the null hypothesis is false, but the test of significance fails to reject it. In general, if we decrease the probability of a Type I error by choosing α small, we increase the probability of a Type II error; that is, we are less likely to reject the null hypothesis when it is false. By choosing α large, we decrease the probability of a Type II error, but we increase the probability of a Type I error.

11.7.3 Two- and One-Sided Tests

In testing the significance of the difference between the two means for the data of Table 11.1, we decided to reject the null hypothesis $\mu_1 = \mu_2$ if the test of significance resulted in either $t \geq 2.101$ or $t \leq -2.101$. The areas of rejection in the t distribution with 18 d.f. were the two tails of the distribution with each tail cutting off 0.025 of the total area. The test of significance, in this instance, is described as a *two-tailed* or *two-sided* test with $\alpha = 0.05$. We made a two-sided test because the outcome of the experiment might have resulted in our finding that $\bar{X}_1 - \bar{X}_2$ was significantly greater than 0 or that $\bar{X}_1 - \bar{X}_2$ was significantly smaller than 0 and we were interested in both of these possibilities.

In some experiments we may be interested only in the possibility that $\bar{X}_1 - \bar{X}_2$ will be significantly larger than 0 and have no interest at all in the possibility that $\bar{X}_1 - \bar{X}_2$ might be significantly smaller than 0. For example, $\bar{X}_1$ may be the mean for an experimental group that receives a treatment which we anticipate will improve performance on a dependent variable and $\bar{X}_2$ may be the mean for a control group that does not receive the treatment. In this instance, we are ordinarily interested only in the possibility that $\bar{X}_1$, the mean for the experimental group, will be significantly larger than $\bar{X}_2$, the mean for the control group.

If this had been the case for the data of Table 11.1, our test of significance would have been

$$t = \frac{\bar{X}_1 - \bar{X}_2}{s_{\bar{X}_1 - \bar{X}_2}}$$

and we would have rejected the null hypothesis, with $\alpha = 0.05$, only if $\bar{X}_1 - \bar{X}_2$ was positive and if t was equal to or greater than 1.734. This is the value of t with 18 d.f. that cuts off 0.05 of the total area in the *right* tail of the t distribution. This test would be described as a *one-tailed* or *one-sided* test of significance because the area of rejection in the t distribution is represented by only one tail or side of the distribution.

One-sided tests are commonly used in discrimination experiments in which we wish to determine if a subject's performance is significantly better than that to be expected on the basis of chance and in which we have no interest at all in the possibility that his performance may be significantly worse than chance. For example, if the probability of a correct response on a single trial is P, under the hypothesis that a subject is responding by chance and if a subject is given n independent trials, then only those values of p greater than P would provide evidence that the subject is responding better than chance.

If the test of significance is made in terms of the standard normal distribution, we would have

$$Z = \frac{p - P}{\sigma_p}$$

and we would reject the hypothesis that the subject is responding by chance, with $\alpha = 0.05$, only if Z is equal to or greater than 1.645. The value of $Z = 1.645$ is the value that cuts off 0.05 of the total area in the right tail of the standard normal distribution and the test is a one-sided test of significance.

11.7.4 THE POWER OF A TEST OF SIGNIFICANCE

In general, if we hold α, the significance level constant, we can decrease the probability of a Type II error by increasing the sample size. We can illustrate this with respect to the experimental data of Table 11.1. In this experiment we have $\bar{X}_1 - \bar{X}_2 = 1.00$ and, as an estimate of the common population variance, we have

$$s^2 = \frac{20 + 20}{10 + 10 - 2} = 2.222$$

Now $\bar{X}_1 - \bar{X}_2$ and s^2, in this experiment, are unbiased estimates of $\mu_1 - \mu_2$ and σ^2, respectively, and we shall assume that if the experiment is repeated we would obtain the same values for $\bar{X}_1 - \bar{X}_2$ and s^2. In our repetition of the experiment, however, we assign twenty subjects to each treatment group instead of ten. Then, using (11.8), we would have, as the standard error of the difference between the means,

$$s_{\bar{X}_1 - \bar{X}_2} = \sqrt{\frac{(2)(2.222)}{20}} = 0.47$$

The test of significance of the difference between the two means would, in this instance, result in

$$t = \frac{\bar{X}_1 - \bar{X}_2}{s_{\bar{X}_1 - \bar{X}_2}} = \frac{1.00}{0.47} = 2.13$$

This t is not only larger than the t of 1.49 we obtained in the original experiment, but it is also based on a larger number of degrees of freedom.

For 38 d.f., an absolute value of t equal to or greater than 2.025 has a probability of 0.05 of occurring when the null hypothesis is true. Because our obtained $t = 2.13$ is greater than the tabled value significant at the 0.05 level of significance for a two-sided test, we would, in this instance, reject the null hypothesis, whereas in the original experiment with the same significance level we did not. In other words, if the null hypothesis $\mu_1 = \mu_2$ is false, then the original experiment would have resulted in our making a Type II error whereas in the repetition with twenty subjects assigned to each treatment, we would not make this error.

If a null hypothesis is, in fact, false, and should be rejected, we define the power of a test of significance as

$$\text{Power} = 1 - \text{P(Type II error)} \qquad (11.10)$$

This is equivalent to saying that the power of a test is the probability that the null hypothesis will be rejected when it should be, that is, when it is false. In general, increasing the number of observations in an experiment will result in increasing the power of a test of significance.

11.8 Confidence Limits for a Difference between Two Means

In the same manner in which we found a 95 percent confidence interval for a single mean μ, we can find a 95 percent confidence interval for the difference between two means, $\mu_1 - \mu_2$. In this case we have

$$\frac{(\bar{X}_1 - \bar{X}_2) - c_1}{s_{\bar{X}_1 - \bar{X}_2}} = t_{0.975} \quad \text{and} \quad \frac{(\bar{X}_1 - \bar{X}_2) - c_2}{s_{\bar{X}_1 - \bar{X}_2}} = t_{0.025}$$

For the data of Table 11.1 we have $\bar{X}_1 - \bar{X}_2 = 4.00 - 3.00 = 1.00$ and $s_{\bar{X}_1 - \bar{X}_2} = 0.67$. With $n_1 + n_2 - 2 = 10 + 10 - 2 = 18$ d.f., the critical values of t are 2.101 and -2.101, respectively. Substituting in the above equations and solving for the confidence limits, c_1 and c_2, we obtain

$$c_1 = 1.00 - (2.101)\,(0.67) = 1.00 - 1.41 = -0.41$$

and

$$c_2 = 1.00 + (2.101)\,(0.67) = 1.00 + 1.41 = 2.41$$

as the lower and upper confidence limits, respectively. The 95 percent confidence interval for $\mu_1 - \mu_2$ is interpreted in the same way as the 95 percent confidence interval for μ.

Note that if the difference between the two sample means had been $\bar{X}_1 - \bar{X}_2 = -1.00$ instead of $\bar{X}_1 - \bar{X}_2 = 1.00$, then we would have had $c_1 = -2.41$ for the lower confidence limit and $c_2 = 0.41$ for the upper confidence limit.

If we make a two-sided test of the null hypothesis $\mu_1 - \mu_2 = 0$ and if, with $\alpha = 0.05$, the obtained value of *t* for the experimental data is such that we fail to reject the null hypothesis, then it will also be true that a 95 percent confidence interval for the same data will include the value of 0. On the other hand, if the null hypothesis is rejected, then it will also be true that the value of 0 will fall outside the 95 percent confidence interval. In other words, a 95 percent confidence interval not only provides us with the same information as the two-sided test of significance but, in addition, provides us with some indication of the limits of the magnitude of the difference between μ_1 and μ_2.

11.9 Why We Should Have $n_1 = n_2$

In illustrating the *t* test for the difference between the means of two independent random samples, we assigned the same number of subjects to each of the two treatment groups. The *t* test can, of course, be used in evaluating the difference between two means when n_1 is not equal to n_2. There are, however, two good reasons for having the same number of subjects in each of the two treatment groups. One of these reasons is that if it is true that $\sigma_1^2 = \sigma_2^2$, then the standard error of the difference will take its minimum value when n_1 is equal to n_2. The second reason is that if it is not true that $\sigma_1^2 = \sigma_2^2$, then we have a violation of one of the assumptions of the *t* test. In this case, it can be shown that if n_1 is equal to n_2, then the *t* test is relatively little influenced by the fact that σ_1^2 is not equal to σ_2^2. In general, therefore, it is advisable to have the same number of subjects or observations for each of the two treatment groups.

11.10 Difference between Two Means: Paired Observations

In some experiments, a single group of subjects is tested first under one treatment and then again under another treatment, so that we have two observations, X_1 and X_2, for each subject. Our primary interest is in the differences between these paired observations or in the values of $D = X_1 - X_2$.

We note that if we sum the values of D, we obtain

$$\Sigma D = \Sigma X_1 - \Sigma X_2$$

If we divide both sides of the above expression by n, the number of differences, we see that the mean of the differences is equal to the difference between the two means or

$$\bar{D} = \bar{X}_1 - \bar{X}_2$$

The estimate of the population variance of the values of D will be given by

$$s_D{}^2 = \frac{\Sigma(D-\bar{D})^2}{n-1} \tag{11.11}$$

where n is again the number of differences or values of D. Then for the variance of the mean difference, $\bar{D}$, we have

$$s_{\bar{D}}{}^2 = \frac{s_D{}^2}{n} \tag{11.12}$$

or

$$s_{\bar{D}}{}^2 = \frac{\Sigma(D-\bar{D})^2}{n(n-1)} \tag{11.13}$$

and the standard error of $\bar{D}$ will be

$$s_{\bar{D}} = \sqrt{\frac{\Sigma(D-\bar{D})^2}{n(n-1)}} \tag{11.14}$$

We illustrate the necessary calculations for the data of Table 11.2. For the values of D, we have

$$\Sigma(D-\bar{D})^2 = \Sigma D^2 - \frac{(\Sigma D)^2}{n} \tag{11.15}$$

or

$$\Sigma(D-\bar{D})^2 = 36 - \frac{(10)^2}{10} = 26$$

Then

$$s_{\bar{D}} = \sqrt{\frac{26}{10(10-1)}} = 0.54.$$

Table **11.2** — Values of a Variable X Obtained from a Single Sample of 10 Subjects with Each Subject Tested Twice

	(1) X_1	(2) X_2	(3) D	(4) D^2
	2	1	1	1
	5	2	3	9
	2	4	−2	4
	3	3	0	0
	7	4	3	9
	3	2	1	1
	5	6	−1	1
	4	3	1	1
	5	4	1	1
	4	1	3	9
Σ	40	30	10	36

For the same data, we have $\bar{D} = \bar{X}_1 - \bar{X}_2 = 4.0 - 3.0 = 1.0$.
  We now define t as

$$t = \frac{\bar{D} - \mu_D}{s_{\bar{D}}} \tag{11.16}$$

and this t will have $n - 1$ d.f., where n is the number of values of D. We wish to test the null hypothesis that $\mu_1 = \mu_2$. If this hypothesis is true, then $\mu_D = \mu_1 - \mu_2$ will be equal to zero. For our example we have $\bar{D} = 1.0$, $s_{\bar{D}} = 0.54$, and consequently

$$t = \frac{1.0 - 0}{0.54} = 1.85$$

with 9 d.f. From the table of t, we find that with 9 d.f. the probability of obtaining t equal to or greater than 1.85 is approximately 0.05 and the probability of obtaining an absolute value of t equal to or greater than 1.85 is, therefore, approximately 0.10, if the null hypothesis that μ_D is equal to zero is true.

REVIEW OF TERMS

confidence coefficient	power of a test
confidence interval	standard error
confidence limits	test of significance
degrees of freedom	Type I error
level of significance	Type II error
null hypothesis	

REVIEW OF SYMBOLS

s^2	s_D^2	α
D	$s_{\bar{X}}^2$	d.f.
c_1	t	σ^2
c_2	μ	n
Z	$s_{\bar{X}_1 - \bar{X}_2}$	s^2/n

QUESTIONS AND PROBLEMS

11.1   We have a random sample of $n = 25$ observations drawn from a normally distributed population with unknown variance and unknown mean. For the sample, we have $s = 10.00$ and $\bar{X} = 18.00$. (a) What is the value of the standard error of the mean? (b) Establish a 95 percent confidence interval for μ.

11.2   Select a point of random entry into the table of random numbers. Take the numbers in two adjacent columns of the table and record the first 10 unlike numbers from 00 to 19. Let these numbers correspond to the subjects with the same numbers in the table below. Call these subjects Sample 1 and the remaining subjects Sample 2. (a) Use the t test to test the null hypothesis that Sample 1 and Sample 2 have been drawn from the same population. (b) Establish a 95 percent confidence interval for μ for both samples. (c) Assume that the 20 observations constitute a population. Find the mean of this population. (d) Do the two confidence intervals established in (b) contain μ?

Subject	X	Subject	X	Subject	X	Subject	X
00	3	05	3	10	1	15	2
01	4	06	5	11	2	16	1
02	3	07	3	12	3	17	4
03	3	08	2	13	4	18	3
04	4	09	5	14	2	19	3

11.3 Twenty subjects were given an intelligence test and paired on the basis of their scores. One member of each pair was then assigned at random to Treatment 1 and the other to Treatment 2. Under these experimental conditions, a measure on a dependent variable X was obtained for each subject. The results are shown below. The observations in this example are *paired*. Find the values of $D = X_1 - X_2$ and test the null hypothesis that $\mu_D = \mu_1 - \mu_2 = 0$.

Pair	Treatment 1	Treatment 2
1	10	7
2	5	3
3	6	7
4	7	5
5	10	8
6	6	4
7	7	5
8	8	2
9	6	3
10	5	6

11.4 Determine a point of random entry into the table of random numbers. Read down the column at the point of entry. Copy the first 20 numbers you find and call this Sample 1. Copy the next 20 numbers and call this Sample 2. Assume that the numbers you have copied represent values of a variable X. (a) Test the null hypothesis that the two samples have been drawn from a common population. (b) Establish a 95 percent confidence interval for both samples. (c) The numbers $0, 1, 2, \ldots, 9$ in the table of random numbers are assumed to be random and, consequently, we may assign a probability of 0.1 to each number. Given the possible values of X and assuming that each value has a probability of 0.1, you can find μ. Do so. (d) Do the two confidence intervals established for the two samples contain μ?

11.5   A control and an experimental group are to be compared. Performance scores of the subjects are given below. Test the null hypothesis that $\mu_1 = \mu_2$.

Control	Experimental
10	7
5	3
6	5
7	7
10	8
6	4
7	5
8	6
6	3
5	2

11.6   We have a random sample of $n = 49$ observations with $\bar{X} = 30$ and $s = 10.5$. Determine a 95 percent confidence interval for μ.

11.7   What is meant by the power of a test of a statistical hypothesis? Other things being equal, if we make the level of significance, α, smaller, will this increase or decrease the power of a test? Explain why.

11.8   We have a population consisting of the N ranks, $1, 2, 3, \ldots, 25$. A random sample of $n = 5$ is drawn from this population *without* replacement. Would it make sense to use the data for the sample of $n = 5$ to find a 95 percent confidence interval for μ? Explain why. If the experiment were repeated 500 times, how many of the independently established 95 per cent confidence intervals would you expect to contain μ?

11.9   In testing the significance of the difference between the means of two independent random samples by the t test, is it desirable to have $n_1 = n_2$? Explain why.

11.10   We make a t test of the significance of the difference between the means of two independent random samples. In evaluating the results of the test is it possible for us to make both a Type I and a Type II error? Explain why.

11.11   What is the crucial distinction between a Z test and a t test?

11.12   We have pre- and posttreatment scores for the same $n = 5$ subjects. Test the null hypothesis that $\mu_D = \mu_1 - \mu_2 = 0$.

Post	Pre
6	4
7	3
9	4
5	2
3	4

11.13 We have a random sample of $n = 25$ observations with $\bar{X} = 73$ and $s = 8$. (a) What is the value of the standard error of the mean? (b) Find a 95 percent confidence interval for μ.

11.14 We have randomly assigned $n = 5$ subjects to each of two treatments. On a dependent variable of interest we have the following measures for the subjects:

Treatment 1	Treatment 2
8	3
4	1
7	5
6	2
5	4

Test the null hypothesis that $\mu_1 = \mu_2$.

11.15 In an experiment with $n = 10$ subjects assigned to each of two treatments, we have $s^2 = 20.0$. (a) What is the value of $s_{\bar{X}_1 - \bar{X}_2}$? (b) If $\bar{X}_1 - \bar{X}_2 = 3.0$, find the value of t for the test of significance of the null hypothesis $\mu_1 = \mu_2$. (c) Assume that s^2 and $\bar{X}_1 - \bar{X}_2$ remain the same in a repetition of the experiment. How many subjects should we assign to each treatment group if we want $s_{\bar{X}_1 - \bar{X}_2}$ to be equal to 1.0? (d) What would be the value of t in the repetition of the experiment?

12

Introduction to the Analysis of Variance

12.1 Introduction

Let us suppose that we have three experimental treatments of interest and that subjects have been randomly assigned to the treatments with the restriction that we have an equal number of subjects for each treatment. The resulting observations obtained under each of the three treatments are shown in Table 12.1 If we have three or more sets of observations, we can test the *null hypothesis* that the observations are independently and randomly drawn from identical normally distributed populations by means of the F distribution. The method of analysis we shall use is known as the *analysis of variance*. We shall illustrate the necessary calculations in terms of the three sets of observations shown in Table 12.1 and then examine the nature of the test of significance.

Table **12.1** — Values of a Variable X Obtained in Each of Three Samples

	Treatment 1		Treatment 2		Treatment 3	
	X_1	$X_1{}^2$	X_2	$X_2{}^2$	X_3	$X_3{}^2$
	12	144	9	81	6	36
	11	121	9	81	7	49
	10	100	7	49	3	9
	10	100	6	36	2	4
	7	49	4	16	2	4
Σ	50	514	35	263	20	102

12.2 Sums of Squares

We first determine the total sum of squares for the $n = 15$ observations, ignoring the fact that they have been classified in three groups of five each, corresponding to the three treatments. This sum of squares will be given by

$$\Sigma(X - \bar{X})^2 = \Sigma X^2 - \frac{(\Sigma X)^2}{n} \tag{12.1}$$

where $n = n_1 + n_2 + n_3$, or the total number of observations and $\bar{X}$ is the mean of all n observations. For the data of Table 12.1, we have $\Sigma X^2 = 514 + 263 + 102 = 879$, $\Sigma X = 50 + 35 + 20 = 105$, and

$$\Sigma(X - \bar{X})^2 = 879 - \frac{(105)^2}{15} = 144$$

We then find a sum of squares that we shall call the *treatment sum of squares*. In general, if we have k treatments with $n_1, n_2, \ldots, n_k$ observations for the respective treatment groups, then the treatment sum of squares will be given by

$$\Sigma n_k(\bar{X}_k - \bar{X})^2 = \frac{(\Sigma X_1)^2}{n_1} + \frac{(\Sigma X_2)^2}{n_2} + \cdots + \frac{(\Sigma X_k)^2}{n_k} - \frac{(\Sigma X)^2}{n} \tag{12.2}$$

where $\Sigma X_1, \Sigma X_2, \ldots, \Sigma X_k$ are sums for the respective treatment groups. For the data of Table 12.1, we have

$$\Sigma n_k(\bar{X}_k - \bar{X})^2 = \frac{(50)^2}{5} + \frac{(35)^2}{5} + \frac{(20)^2}{5} - \frac{(105)^2}{15} = 90$$

If we subtract the treatment sum of squares from the total sum of squares, we obtain a sum of squares that we shall call the *within treatment sum of squares*. Thus

$$\text{Within} = \text{Total} - \text{Treatment} \tag{12.3}$$

The within treatment sum of squares is a pooled sum of squares based on the variation of the measures within each treatment group about the respective means of the groups. For example, if we consider each group of observations in Table 12.1 separately, we would have

$$\Sigma(X_1 - \bar{X}_1)^2 = 514 - \frac{(50)^2}{5} = 14$$

$$\Sigma(X_2 - \bar{X}_2)^2 = 263 - \frac{(35)^2}{5} = 18$$

$$\Sigma(X_3 - \bar{X}_3)^2 = 102 - \frac{(20)^2}{5} = 22$$

and the sum of these three sums of squares is 54 or the within treatment sum of squares.

12.3 The Test of Significance

The results of the calculations are summarized in Table 12.2. Each of the sums of squares we have calculated has associated with it a specified number of degrees of freedom. For the total sum of squares, the number of degrees of freedom is equal to $n-1$. For the sum of squares within the first group, we have n_1-1 d.f.; for that within the second group, we have n_2-1 d.f.; and for the third group, n_3-1 d.f. Because we have pooled these three sums of squares to obtain the within treatment sum of squares, we also pool the degrees of freedom. In general, the within treatment sum of squares will have $n-k$ d.f., where n is the total number of observations and k is the number of different treatments or sets of observations. For the treatment sum of squares, we will have $k-1$ d.f., where k is the number of treatment groups.

Table **12.2** — Summary Table of the Analysis of Variance for the Observations in Table 12.1

(1) Source of Variation	(2) Sum of Squares	(3) d.f.	(4) Mean Square
Treatments	90	2	45.0
Within treatments	54	12	4.5
Total	144	14	

If we divide the treatment sum of squares by its degrees of freedom, and the within treatment sum of squares by its degrees of freedom, we obtain two variance estimates that are called *mean squares*. These are shown in column (4) of Table 12.2. We then define F as

$$F = \frac{MS_T}{MS_W} \tag{12.4}$$

where MS_T is the treatment mean square and MS_W is the within treatment mean square. The F ratio defined by (12.4) will have $k-1$ d.f. for the numerator and $n-k$ d.f. for the denominator.[1]

For the problem under discussion, we have

$$F = \frac{45.0}{4.5} = 10.0$$

with 2 and 12 d.f. To evaluate the probability of this F, when the null hypothesis is true, we enter the column of the table of F, Table V in the

---

[1]In the previous chapter, we used t to test the significance of the difference between the means of $k=2$ independent random samples. In the case of $k=2$ samples, the F ratio defined by (12.4) will be equal to t^2.

Appendix, with the 2 d.f. corresponding to the numerator, and run down this column until we find the row entry corresponding to the 12 d.f. of the denominator. The tabled values of F are those that would be expected to be equaled or exceeded with a probability of 0.05 and 0.01 when the null hypothesis is true. The 0.05 values are given in lightface type and the 0.01 values in boldface type,

For 2 and 12 d.f., we see that F equal to or greater than 6.93 would be expected to occur under the null hypothesis with a probability of 0.01. The probability associated with $F = 10.0$ must, therefore, be considerably less than 0.01. If we have chosen α equal to 0.05 or 0.01, the null hypothesis would be rejected.

Assuming the null hypothesis is true, the distribution of F is known exactly. As is obvious from the table of F in the Appendix, we have a different F distribution for each pair of degrees of freedom, one member of the pair corresponding to the numerator of the F ratio and the other to the denominator.

12.4 The Within Treatment Mean Square

If the null hypothesis is true, then each of the variances obtained from each of the treatment groups is an unbiased estimate of the same population variance. As in the case of the t test, where we had only two estimates of the same population variance, we pool the sums of squares within each treatment group and corresponding degrees of freedom to obtain an estimate of the common population variance. Thus, if we have k treatments, then

$$s^2 = \frac{(n_1 - 1)s_1^2 + (n_2 - 1)s_2^2 + \cdots + (n_k - 1)s_k^2}{(n_1 - 1) + (n_2 - 1) + \cdots + (n_k - 1)}$$

or

$$s^2 = \frac{\Sigma(X_1 - \bar{X}_1)^2 + \Sigma(X_2 - \bar{X}_2)^2 + \cdots + \Sigma(X_k - \bar{X}_k)^2}{n - k} \tag{12.5}$$

where n equals the total number of observations and k the number of treatments. The numerator of (12.5) we have called the within treatment sum of squares, and the denominator is the number of degrees of freedom associated with the within treatment sum of squares. Thus (12.5) defines the within treatment mean square.

12.5 The Treatment Mean Square

Now let us examine the *treatment mean square*. If the null hypothesis is true, each of the k sample means $\bar{X}_1, \bar{X}_2, \ldots, \bar{X}_k$ is an unbiased estimate of the same population mean. Then we can combine the observations from the different treatments to obtain as an estimate of the common population mean

$$\bar{X} = \frac{n_1\bar{X}_1 + n_2\bar{X}_2 + \cdots + n_k\bar{X}_k}{n_1 + n_2 + \cdots + n_k} \tag{12.6}$$

and the value obtained by (12.6) is the mean of the total set of n observations.

We have already found that the variance of the sampling distribution of means of samples drawn from the same population is given by

$$\sigma_{\bar{X}}^2 = \sum P(\bar{X} - \mu)^2 \tag{12.7}$$

We do not know the value of μ. We do, however, have an unbiased estimate of μ, as given by $\bar{X}$, the mean of the combined samples, if the null hypothesis is true. Then the variance given by (12.7) can be estimated by

$$s_{\bar{X}}^2 = \frac{(\bar{X}_1 - \bar{X})^2 + (\bar{X}_2 - \bar{X})^2 + \cdots + (\bar{X}_k - \bar{X})^2}{k-1} \tag{12.8}$$

where k is the number of sample means or treatment groups.

We also know that the variance of the means of random samples of size n drawn from a common population is given by

$$\sigma_{\bar{X}}^2 = \frac{\sigma^2}{n} \tag{12.9}$$

and consequently

$$n\sigma_{\bar{X}}^2 = \sigma^2 \tag{12.10}$$

Then, $ns_{\bar{X}}^2$ should also provide an estimate of the population variance σ^2. Multiplying both sides of (12.8) by n_k, the number of observations in each of the k samples, we have

$$n_k s_{\bar{X}}^2 = \frac{n_k[(\bar{X}_1 - \bar{X})^2 + (\bar{X}_2 - \bar{X})^2 + \cdots + (\bar{X}_k - \bar{X})^2]}{k-1} \tag{12.11}$$

and the right-hand side of (12.11) is also an estimate of the common population variance, if the null hypothesis is true.

The numerator of (12.11) is identical with (12.2) and is the treatment sum of squares. The denominator is the number of degrees of freedom associated with the treatment sum of squares. Thus (12.11) defines the mean square for treatments.

12.6 The Null Hypothesis

The F defined by (12.4) is the treatment mean square divided by the within treatment mean square. If the null hypothesis is true, then both the numerator and denominator of the F ratio are estimates of the same common population variance. Consequently, we should expect the numerator to exceed the denominator only as a result of sampling variation. The values of F given in Table V in the Appendix are those values that would be equaled or exceeded with a probability of 0.05 or 0.01, when the null hypothesis is true. Thus, if we choose $\alpha = 0.05$ and if we obtain a value of F that exceeds the tabled value, we may decide to reject the null hypothesis.

If the obtained value of F has a small probability, then either a rare or improbable event has occurred by chance or the theoretical model, as given by the null hypothesis, is false. The null hypothesis states that the k samples have been randomly and independently drawn from identically distributed normal populations. Examination of the null hypothesis indicates that it consists of four assumptions: (1) the observations are independent and random; (2) the population distributions of X are normal in form; (3) the populations have the same variances; and (4) the populations have the same mean. We try to ensure that (1) is true by randomly assigning subjects to the k treatment groups by using a table of random numbers or some other method of random assignment. It is now well established that the F test is, like the t test, relatively insensitive to departures from normality of distribution and we should not obtain a significant value of F simply because (2) is false.[2] Similarly, it has been shown that the F test is relatively insensitive to differences in variances, provided the same number of observations are present for each treatment group. Therefore, if we have assigned the same number of subjects to each treatment group, we should not obtain a significant value of F simply because the variances are not equal. Thus, it seems that the most reasonable interpretation of a significant value of F is that the population means for the various treatment groups are not all equal to the same value. This, of course, is what we are primarily interested in.

12.7 Tests on the Treatment Means

If the obtained value of F is significant, then we would ordinarily like to make additional tests of significance that will tell us something more about the nature of the differences among the means. We describe a method proposed by Scheffé, called the *S-method*, for evaluating various comparisons among the treatment means.[3] We shall discuss only the case where we have the same number of observations for each treatment.

---

[2] In general, departures from normality of distribution are relatively unimportant, provided we have an adequate number of observations for each of the k treatments. What will constitute an "adequate" number of observations will depend on the nature of the departures from normality. For example, for moderately skewed populations $n \geqslant 10$ observations per treatment should be adequate. For more extreme degrees of skewness this number should be increased. Even with fairly drastic departures from normality, $n \geqslant 25$ observations per treatment should be adequate.

[3] H. Scheffé. A method for judging all contrasts in the analysis of variance. *Biometrika*, 1953, *40*, 87–104.

Consider any two treatment means, say $\bar{X}_1$ and $\bar{X}_2$. Suppose we multiply $\bar{X}_1$ by some number, say a_1, and that we multiply $\bar{X}_2$ by some number a_2, to obtain

$$d = a_1\bar{X}_1 + a_2\bar{X}_2 \tag{12.12}$$

Then d is described as a *comparison* if and only if $\Sigma a = a_1 + a_2 = 0$. In the case of a comparison involving only two means, the obvious values for a_1 and a_2 are 1 and -1, respectively. In this case, then, we have

$$d = \bar{X}_1 - \bar{X}_2$$

and this is simply a comparison involving the difference between $\bar{X}_1$ and $\bar{X}_2$. The numbers $a_1 = 1$ and $a_2 = -1$ are called coefficients of the means.

The *standard error* of a comparison will be given by

$$s_d = \sqrt{\frac{s^2}{n_k}\Sigma a^2} \tag{12.13}$$

where s^2 is the within treatment mean square and n_k is the number of observations on which each treatment mean is based. Then, as a test of significance, we have

$$t = d/s_d \tag{12.14}$$

Let us make all possible comparisons of the differences between two means for the data of Table 12.1. Making these comparisons, we have

$$\bar{X}_1 - \bar{X}_2 = 10 - 7 = 3$$
$$\bar{X}_1 - \bar{X}_3 = 10 - 4 = 6$$
$$\bar{X}_2 - \bar{X}_3 = 7 - 4 = 3$$

We also have, for the data of Table 12.1, $s^2 = 4.5$ and $n_k = 5$ observations for each treatment mean. Because $\Sigma a^2 = 2$ for all of the above comparisons, each comparison will have the same standard error. Thus

$$s_d = \sqrt{\frac{4.5}{5}(2)} = 1.34$$

Testing each comparison for significance, we obtain the following values of t:

$$t = \frac{\bar{X}_1 - \bar{X}_2}{s_d} = \frac{3}{1.34} = 2.24$$

$$t = \frac{\bar{X}_1 - \bar{X}_3}{s_d} = \frac{6}{1.34} = 4.48$$

$$t = \frac{\bar{X}_2 - \bar{X}_3}{s_d} = \frac{3}{1.34} = 2.24$$

To evaluate a t obtained by using the S-method, we find

$$t' = \sqrt{(k-1)F} \tag{12.15}$$

where F is the tabled value of F at the 0.05 or 0.01 level of significance for the degrees of freedom of the analysis of variance. For example, in the problem under discussion, we have $k-1 = 2$ d.f. for the numerator of the F ratio and $k(n_k-1) = 12$ d.f. for the denominator. The tabled value of F for 2 and 12 d.f. at the 0.05 level is 3.88. Then, substituting in (12.15), we have

$$t' = \sqrt{(3-1)(3.88)} = 2.79$$

and $t' = 2.79$ is the standard in terms of which we evaluate all of the t's obtained from the comparisons we have made. The only significant value of t, that is, the only value that exceeds $t' = 2.79$, is that obtained when we compared the difference between $\bar{X}_1$ and $\bar{X}_3$. The other two t's are less than $t' = 2.79$ and, therefore, are not significant by the S-method.

The S-method is not limited to making comparisons between any two means. It can also be used to make comparisons involving three or more means. For example, each of the following is a comparison because, as can readily be determined, the sum of the coefficients of the means in each case is equal to zero:

$$\bar{X}_1 - \frac{1}{2}\bar{X}_2 - \frac{1}{2}\bar{X}_3 = \bar{X}_1 - \frac{\bar{X}_2 + \bar{X}_3}{2}$$

$$\frac{1}{2}\bar{X}_1 - \bar{X}_2 + \frac{1}{2}\bar{X}_3 = \frac{\bar{X}_1 + \bar{X}_3}{2} - \bar{X}_2$$

$$\frac{1}{2}\bar{X}_1 + \frac{1}{2}\bar{X}_2 - \bar{X}_3 = \frac{\bar{X}_1 + \bar{X}_2}{2} - \bar{X}_3$$

The first comparison above is obviously a comparison of the difference between $\bar{X}_1$ and the average value of $\bar{X}_2$ and $\bar{X}_3$. The second is a comparison of the difference between the average value of $\bar{X}_1$ and $\bar{X}_3$ and the value of $\bar{X}_2$. The third is a comparison of the difference between the average value of $\bar{X}_1$ and $\bar{X}_2$ and the value of $\bar{X}_3$. Substituting with the values of $\bar{X}_1$, $\bar{X}_2$, and $\bar{X}_3$ for the data of Table 12.1, we have

$$10 - \frac{7+4}{2} = 4.5$$

$$\frac{10+4}{2} - 7 = 0$$

$$\frac{10+7}{2} - 4 = 4.5$$

We note that for each of the above comparisons, we have

$$\Sigma a^2 = a_1{}^2 + a_2{}^2 + a_3{}^2 = \frac{3}{2}$$

and, therefore, the standard error for each of the above comparisons will be

$$s_d = \sqrt{\frac{4.5}{5}\left(\frac{3}{2}\right)} = 1.16$$

Testing the first comparison for significance, we obtain

$$t = \frac{4.5}{1.16} = 3.88$$

and $t = 3.88$ is greater than $t' = 2.79$ and may be judged significant. For the third comparison we also have $t = 3.88$ and this is also a significant value. We may conclude, therefore, that $\bar{X}_1$ differs significantly from the average value of $\bar{X}_2$ and $\bar{X}_3$. Similarly, the value of $\bar{X}_3$ differs significantly from the average value of $\bar{X}_1$ and $\bar{X}_2$.

The S-method of testing comparisons is such that if $F = MS_T/MS_W$ is significant with $\alpha = 0.05$, then at least one of the comparisons that can be made on a set of k means will also be significant. There may, of course, be more than one.

12.8 The Analysis of Variance and Experimental Design

The analysis of variance plays an important role in the design of experiments and in the analysis of experimental data. The experimental design that we have illustrated in this chapter is called a *randomized group design*. The essential characteristic of this design is that subjects are randomly assigned to the experimental treatments (or vice versa). For example, if we have eight subjects and we wish to test four under Treatment 1 and four under Treatment 2, the randomization process gives every possible group of four subjects selected from a set of eight an equal probability of being assigned to Treatment 1 and Treatment 2.

A variety of other useful experimental designs have been developed that involve the analysis of variance. In a *randomized block design*, subjects are first divided into groups such that, within each group, the subjects are relatively homogeneous with respect to some selected variable. The variable selected for grouping the subjects is one that is believed to be related to the measures to be obtained on the dependent variable. In a learning experiment, for example, subjects might be grouped into blocks such that the subjects within each block have comparable IQ's. The number of subjects in a given block would be equal to the number of treatments, and one subject in each block would be randomly assigned to each treatment. In the analysis of variance, a sum of squares associated with differences between blocks (levels of IQ) would be isolated. If the blocks constitute an important source of variation, it is expected that the mean square used in the denominator of the F ratio will be smaller than the mean square obtained from a randomized group design.

Another type of experiment frequently used in research is a *factorial experiment*. In a factorial experiment, two or more independent variables, each varied in two or more ways, are studied in all possible combinations within the framework of a single experiment. We might have, for example, a variable A that is varied in two ways, A_1 and A_2, and another variable B that is also varied in two ways, B_1 and B_2. The possible combinations of these would give A_1B_1, A_1B_2, A_2B_1, and A_2B_2. Each of the four combinations would provide one treatment or experimental condition. Subjects would then be assigned at random to each treatment in the same manner as in a randomized group design. The value of the factorial experiment is that it permits an evaluation of the difference between the means for A_1 and A_2, and also an evaluation of the difference between the means for B_1 and B_2, as well as various other comparisons of interest.

REVIEW OF TERMS

analysis of variance	randomized group design
comparison	S-method
degrees of freedom	total sum of squares
factorial experiment	treatment mean square
mean square	treatment sum of squares
null hypothesis	within treatment mean square
randomized block design	within treatment sum of squares

REVIEW OF SYMBOLS

$\Sigma(X-\bar{X})^2$	α	F
$\Sigma n_k(\bar{X}_k-\bar{X})^2$	d	MS_T
n	Σa^2	MS_W
k	s_d	s^2
n_k	t'	$s_{\bar{X}}^2$

QUESTIONS AND PROBLEMS

12.1 Subjects were assigned at random to four treatments and the measures on a dependent variable X of interest for the subjects in each treatment are given below.

Treatments			
1	*2*	*3*	*4*
8	9	5	6
5	4	3	1
6	8	7	1
8	4	5	6
9	3	3	5
10	6	1	4
9	7	5	3
7	6	4	6
8	7	3	4
10	6	4	4

Use the analysis of variance to determine if the treatment means differ significantly.

12.2   Consider only Treatments 1 and 2. (a) Test the significance of the difference between the two treatment means using the t test. (b) Now use the analysis of variance to test the significance of the treatment mean square for the two groups. You should find that the value of F is equal to the value of t^2 that you obtained in (a).

12.3   For the case of two groups with $n_1 = n_2 = n$, the treatment sum of squares will be given by

$$\frac{(\Sigma X_1 - \Sigma X_2)^2}{2n}$$

Substitute in the above expression with the treatment sums for Treatments 1 and 2 and see that the value you obtain is equal to the treatment sum of squares of Problem 12.2.

12.4   State, in your own words, the nature of the variation measured by (a) the total sum of squares, (b) the treatment sum of squares, and (c) the within treatment sum of squares.

12.5   What assumptions are involved in the null hypothesis tested by F? Which of these assumptions can be relaxed and under what conditions?

12.6   Make the following comparisons on the treatment means of Problem 12.1 and test each for significance using the S-method.

(a) $\bar{X}_1 - \frac{1}{3}(\bar{X}_2 + \bar{X}_3 + \bar{X}_4)$

(b) $\bar{X}_2 - \frac{1}{2}(\bar{X}_3 + \bar{X}_4)$

(c) $\bar{X}_3 - \bar{X}_4$

12.7   Forty subjects were assigned at random to five treatments. Measures on a dependent variable X for each treatment group are given below.

Treatments

1	2	3	4	5
18	18	4	7	9
13	9	13	3	16
21	15	11	11	26
14	25	11	11	21
25	14	15	7	18
14	6	15	13	11
7	12	11	10	14
20	9	12	10	13

Use the analysis of variance to determine whether the treatment means differ significantly.

12.8 Assume that for a control condition we obtain a given mean and variance on a dependent variable X of interest. (a) Can you think of a treatment or experimental condition that might serve to increase the variance of X without changing the mean? Explain your answer. (b) Can you think of a treatment or experimental condition that might serve to decrease the variance of X without changing the mean? Explain your answer.

12.9 What is the null hypothesis tested by the F test in the analysis of variance? Does this null hypothesis differ in any way from the null hypothesis tested by the t test?

12.10 Ten subjects are randomly divided into two groups of five subjects each. One group receives Treatment 1 and the other Treatment 2. Measures on a dependent variable of interest obtained after the administration of the treatments are given below.

Treatment 1	Treatment 2
8	3
6	5
5	1
7	4
4	2

(a) Test the null hypothesis that $\mu_1 = \mu_2$ by means of the t test. (b) For the same data calculate $F = MS_T/MS_W$. (c) Is t^2 equal to the value of F?

12.11 Twelve subjects are randomly divided into $k = 4$ groups of $n = 3$ subjects each. Each group is randomly assigned to a different treatment. Measures on a dependent variable of interest obtained after the administration of the treatments are given below.

Treatments

1	2	3	4
4	3	1	8
6	5	3	10
2	7	5	6

Calculate $F = MS_T/MS_W$.

12.12 We have five treatments with five subjects randomly assigned to each treatment. The data are given below. Complete the analysis of variance.

Treatments

1	2	3	4	5
3	3	7	6	7
4	4	6	7	8
1	2	5	8	9
2	5	3	5	5
5	6	4	4	6

12.13 For $k = 2$ treatments with $n = 5$ subjects randomly assigned to each treatment, the outcomes of an experiment are as given below. Complete the analysis of variance.

Treatments

1	2
4	4
2	10
6	8
10	6
8	12

12.14 Calculate t for a test of significance of the difference between the means for Problem 12.13.

13

The χ^2 Test

13.1 Introduction

When we had a variable such that the observations could take only the value of $X = 1$ or $X = 0$, we found that the random sampling distribution of p (and T) for samples of size n was completely determined by P, the proportion of observations in the population with $X = 1$ or, in other words, the probability of $X = 1$. We assumed the sampling distribution to be normal in form if both nP and nQ were equal to or greater than 5, and it was thus possible to transform any observed value of p (or T) into a standard normal deviate Z. By means of the table of the standard normal curve, we then determined the probability of obtaining a value of p (or T) equal to or greater than the observed value. If this probability was small, we rejected the null hypothesis. Because, in the problem considered, $P + Q = 1.00$, if we reject the null hypothesis concerning P, we are, at the same time, rejecting a null hypothesis concerning Q.

13.2 A Single Sample with Several Categories

In the problem to be considered now, we have a variable such that the observations can take any one of three or more values. The values of the variable do not need to be ordered. Suppose, for example, we have three brands of frozen orange juice, and subjects are asked to taste each one and to indicate which of the three they prefer. The observations consist of the responses or preferences expressed by $n = 60$ students, and each observation can be classified according to the brand preferred. We identify the brands by the letters A, B, and C, and the observations are classified

Table **13.1** — Distribution of Preferences for Each of Three Brands in a Sample of 60 Observations and Calculation of χ^2

(1) Brand	(2) f	(3) F	(4) $(f - F)$	(5) $(f - F)^2$	(6) $(f - F)^2/F$
A	16	20	-4	16	0.8
B	28	20	8	64	3.2
C	16	20	-4	16	0.8
Σ	60	60	0		4.8

accordingly. We form a frequency distribution, with f_A, f_B, and f_C corresponding to the frequencies in the respective classes, as shown in Table 13.1.

Our interest in the three frequencies given in Table 13.1 is, we assume, to determine whether they are in accord with some expectation we have formulated prior to making the observations. This expectation provides the motivation for making the observations and corresponds to a null hypothesis we wish to test. If this were not our interest, but we were merely curious as to the nature of the distribution, the work would be completed with the making of the frequency distribution and with perhaps some graphic representation of the distribution. We shall assume, however, that this is not the case, and that our interest in the three frequencies is in terms of a null hypothesis we formulated prior to making the observations. We wish to determine whether the observations are in accord with this null hypothesis.

We might have argued in this way: We do not know that there are any differences in the three brands such that they will not be equally preferred. If there are differences, we do not know which, if any, of the brands may be the preferred one. So, let us assume that there are no differences in the brands. Then a student's preference should be a matter of chance. Because there are three brands and only one can be chosen, the probability of any one being chosen is 1/3. Under this null hypothesis, we have $P_A = P_B = P_C = 1/3$. This is the null hypothesis we wish to test.

In a random sample of n observations, the theoretical frequencies F for each of the classes will be given by multiplying each of the values of P by n. Thus

$$F = nP \tag{13.1}$$

We note that the values of P are assigned in such a way that $\Sigma P = 1.00$ and therefore $\Sigma F = n$.

The theoretical frequencies for the data of Table 13.1, established by the null hypothesis, are given in column (3). The test of significance is accomplished by calculating

$$\chi^2 = \Sigma \frac{(f - F)^2}{F} \tag{13.2}$$

where the summation is over the number of classes. The values of $(f-F)^2$ and $(f-F)^2/F$ are shown in columns (5) and (6), respectively. Summing the entries in column (6), we obtain 4.8 as the value of χ^2.

We can determine the probability of obtaining χ^2 equal to or greater than 4.8, when the null hypothesis is true, by means of Table VI in the Appendix. This is a table of χ^2 and to use the table we must enter it with the number of degrees of freedom associated with χ^2. The number of degrees of freedom in the present problem is determined by the fact that only two of the frequencies are free to vary for any given n. That is, if we know any two of the frequencies, the third can be determined for any fixed n by subtraction. Thus, we say we have only 2 d.f. for the data of Table 13.1.

If we enter the table of χ^2 with 2 d.f., we find that the probability of obtaining a value of χ^2 as large as or larger than 4.605 is 0.10 and the probability of obtaining a value of χ^2 as large as or larger than 5.991 is 0.05. Because the observed value of $\chi^2 = 4.8$ falls close to the tabled value of 4.605, the probability of the obtained value is only slightly less than 0.10. If we had chosen α, the significance level of the test, as 0.05, then we would not reject the null hypothesis. On the other hand, if we had obtained a value of χ^2 equal to or greater than 5.991, and with $\alpha = 0.05$, the null hypothesis would have been rejected. To reject the null hypothesis, in this instance, is to say that we do not believe the observed frequencies f are in accord with the theoretical frequencies F. Because the values of F, in turn, were determined by the values of P, rejection of the null hypothesis is a rejection of the values of P we assigned to the various classes. In other words, if we reject the null hypothesis this means that we do not believe that $P_A = P_B = P_C$.

13.3   Importance of a Prior Null Hypothesis

Examination of Table 13.1 indicates that the largest contribution to χ^2 is provided by the discrepancy between f_B and F_B, that is, for Brand B. We use this fact to indicate the importance of formulating a null hypothesis prior to an examination of the data. The null hypothesis determines how the observations are to be evaluated and not the other way around. Suppose, for example, prior to making the observations, a friend claimed that Brand B would be the preferred brand. The observations are now made with the purpose of providing information about our friend's claim. We are no longer interested in the distinction between Brands A and C, because a choice of either of these two brands will provide evidence that Brand B is not the preferred brand. We therefore combine all such choices in a single class that we designate as "not-B." We thus have observations that can be designated as B or not-B. The distribution would now be as shown in Table 13.2.

Table **13.2** — Calculation of χ^2 for the Data of Table 13.1 when Frequencies for Brands A and C Are Combined

(1) Brand	(2) f	(3) F	(4) $(f-F)$	(5) $(f-F)^2$	(6) $(f-F)^2/F$
B	28	20	8	64	3.2
Not-B	32	40	−8	64	1.6
Σ	60	60	0		4.8

For the theoretical values of P, we again assume that the probability of a B choice is 1/3, if the choice is a matter of chance, and the probability of a not-B choice is 2/3. Then the expected frequencies now become $(60)(1/3) = 20$ and $(60)(2/3) = 40$, as shown in column (3) of Table 13.2. The values of $(f-F)^2$ are given in column (5) and $(f-F)^2/F$ in column (6). Summing the entries in column (6), we have $\chi^2 = 4.8$ as before.

To evaluate the observed value of $\chi^2 = 4.8$, we enter the table of χ^2 with the degrees of freedom available. We now note that we have only 1 d.f., because only one of the two frequencies is free to vary for a fixed n. Knowing one of the two, the other can be obtained by subtraction from n. From the table of χ^2 we find that a value of χ^2 equal to 3.841 or larger can be expected to occur with a probability of 0.05, when the null hypothesis is true. If α, the significance level, is 0.05, then the null hypothesis would be rejected, for we have an obtained value of χ^2 equal to 4.8 for which the probability is less than 0.05. Examination of the data of Table 13.2 indicates that Brand B is, in fact, preferred more often than would be expected on the basis of the null hypothesis.

In the first case, with exactly the same data, the null hypothesis was not rejected, whereas in the present instance it is. We have, in essence, two different null hypotheses that we have tested. Each is a perfectly good hypothesis, if formulated prior to making the observations. By waiting until after we have made the observations and then formulating the null hypothesis on the basis of what we observe, we capitalize on chance factors in such a way as to maximize the obtained value of χ^2. It would be more reasonable, and in accord with good experimental procedures, if the null hypothesis originally formulated was $P_A = P_B = P_C$, to regard the large number of choices given to Brand B as suggestive of a new hypothesis to be tested by the collection of new observations.

We may note that, for the analysis of the data in Table 13.2, the problem is comparable to the one described earlier concerning the discrimination between two brands of cigarettes. For the data of Table 13.2 we also have only two classes, a B choice and a not-B choice. Thus, in this instance, the test of significance could also be made in terms of the standard normal deviate Z. When χ^2 has only 1 d.f., then $\chi^2 = Z^2$. The χ^2 distribution, however, enables us to deal with observations that fall into two or more classes. In general, for problems of the kind described, if we have c classes of observations, then χ^2 as obtained from (13.2) will have $c - 1$ d.f.

13.4 Two or More Samples

A sociologist has interviewed a group of $n = 200$ individuals and has classified these individuals into two groups of 100 each. One group he describes as a low-income group and the other as a high-income group. One of the things he has determined in his interview was whether or not each individual owned a television set. We can consider this as a variable, the observations of which can be classified as Yes, indicating ownership, and No, indicating nonownership. We let the number of individuals in the high-income group be designated by n_1 and the number in the low-income group by n_2. Table 13.3 gives the number of owners and nonowners in each of the two groups.

Table **13.3** — Distribution of Yes and No Responses by Income Classification

	No	Yes	Total
High income	10	90	100
Low income	30	70	100
Total	40	160	200

We thus have two sets of observations, and within each set the observations are classified as Yes or No. The sociologist states that what he wants to find out is whether the proportion of individuals in the high-income group who own sets and the proportion of owners in the low-income group differ significantly. For the high-income group we have $p_1 = 90/100 = 0.90$ and for the low-income group we have $p_2 = 70/100 = 0.70$.

Let us assume, as a null hypothesis to be tested, that the two sets of observations have been drawn at random from a common binomial population. If this hypothesis is true, then let us assume that P, the proportion of owners in the population, will be estimated by

$$P = \frac{n_1 p_1 + n_2 p_2}{n_1 + n_2} \qquad (13.3)$$

and, for the problem under consideration, we have

$$P = \frac{100(0.90) + 100(0.70)}{100 + 100} = \frac{160}{200} = 0.80$$

Under the assumption that P, as defined by (13.3), is the proportion of owners in the population, then $n_1 P$ and $n_2 P$ will give the expected frequency of owners in the high-income and low-income groups, respectively.

Table **13.4** — Theoretical or Expected Frequency of Yes and No Responses for the Data of Table 13.3

	No	Yes	Total
High income	20	80	100
Low income	20	80	100
Total	40	160	200

Similarly, n_1Q and n_2Q, where $Q = 1 - P$, will give the expected frequency of nonowners in the high- and low-income groups, respectively.

Table 13.4 gives the theoretical frequencies of Yes's and No's in the two samples, if the null hypothesis is true. What is the null hypothesis in this instance? It is that the two samples are random samples from a common binomial population in which $P = 0.80$ and $Q = 0.20$. The test of significance is made by means of χ^2 as given by (13.2). We have four observed frequencies f, and four theoretical frequencies F. Then, substituting in (13.2), we have

$$\chi^2 = \frac{(10-20)^2}{20} + \frac{(90-80)^2}{80} + \frac{(30-20)^2}{20} + \frac{(70-80)^2}{80} = 12.50$$

We place the restriction on the data of Table 13.3 that the marginal entries must remain the same. Then it is clear that only one of the four cell entries is free to vary because, if we know any one of the cell entries, the remaining three can be obtained by subtraction from the marginal totals. For this example, then, χ^2 will have only 1 d.f.

Let us assume that we have chosen $\alpha = 0.05$. From the table of χ^2 in the Appendix, we find that for 1 d.f., the value of χ^2 significant with probability of 0.05 is 3.841. Because the obtained value of $\chi^2 = 12.50$ exceeds the tabled value, we reject the null hypothesis. We conclude that the samples have not been randomly drawn from a common binomial population with $P = 0.80$. In essence, we conclude that the two sample p's differ significantly.

We have illustrated the calculation of χ^2 for the case of two samples with only two categories in which the observations in the samples may occur. The classification of the observations then results in a 2×2 table, such as Table 13.3. The application of the χ^2 test, however, is not limited to a 2×2 table. We may, for example, have r samples with c categories of classification where r and c are both greater than two. Under these circumstances, χ^2 will have $(r-1)(c-1)$ d.f., where r is the number of rows or samples and c is the number of columns or categories in which the observations are classified.[1] For the 2×2 table this results in 1 d.f.

---

[1] The calculation of χ^2 for a 3×3 table is shown in Chapter 14. The procedures described there can be used for any $r \times c$ table.

13.5 Correction for Discontinuity

The distribution of χ^2 is a theoretical distribution corresponding to a continuous variable such that the distribution can be represented by a curve, with a different curve for differing degrees of freedom. The true probabilities of obtaining the results shown in the examples cited can be obtained by exact methods. By calculating χ^2 and then evaluating this obtained value in terms of the theoretical distribution of χ^2, we obtain an approximation of the true probabilities calculated by the exact methods. When we have but a single degree of freedom, the approximation can be somewhat improved by a *correction for discontinuity*. We make this correction in (13.2) by subtracting 0.5 from each of the *absolute* values of the discrepancies $f - F$, before squaring them. The effect of the correction is to make these deviations smaller and thus to reduce the obtained value of χ^2. The correction is of relatively little importance in large samples and is *not* made when χ^2 has more than 1 d.f.

REVIEW OF TERMS

absolute value   null hypothesis
correction for discontinuity   observed frequency
degrees of freedom   theoretical frequency

REVIEW OF SYMBOLS

F α
n χ^2
P d.f.
Q

QUESTIONS AND PROBLEMS

Note to the Reader:

In some of the problems given below, you are asked to calculate χ^2 with 1 d.f. without a correction for discontinuity and in other problems to calculate the value with a correction for discontinuity. In still other problems, you are asked to calculate χ^2 with 1 d.f. both with and without a correction for discontinuity. This is done intentionally. In some cases, I want you to see the relationship between χ^2 with 1 d.f. and Z^2 without a correction for discontinuity and, in other cases, with a correction for discontinuity. In other cases, I want you to calculate χ^2 with 1 d.f. without a correction for discontinuity and then to ask you what the appropriate correction should be. In still other cases, the number of observations is sufficiently large so that I do not suggest a correction for discontinuity, although it is always appropriate for any χ^2 with 1 d.f. I want you to see all of these possibilities.

13.1   Previous experience with a particular achievement test indicated that for seventh-grade children the ratio of those receiving a passing mark to those failing was 3 : 1. A teacher wishes to determine whether this ratio also holds for sixth-grade children. In a random sample of 100 students drawn from the sixth grade, she finds that 60 pass the test. (a) Use χ^2 to test the hypothesis that the sixth-grade students are a random sample from a binomial population with $P = 0.75$ and $Q = 0.25$. (b) If a correction for discontinuity is made, then instead of using 60 and 40 as the number passing and failing, respectively, what two values should be used?

13.2   One hundred and twenty students were interviewed concerning their preferences about examinations. Of the 120 students, 55 preferred objective examinations, 35 preferred essay examinations, and 30 preferred short-answer examinations. Test the null hypothesis that the three types of examinations are equally preferred.

13.3   The following table shows the distribution of responses of 30 schizoid and 30 manic patients to an item in a personality inventory.

Group	Yes	?	No	Total
Schizoid	18	9	3	30
Manic	6	9	15	30
Total	24	18	18	60

Test the null hypothesis that the schizoid and manic responses are from a common population.

13.4   An experimental group of $n = 80$ rats received a treatment which would presumably increase the probability of a right turn at a choice point in a T maze. A control group of $n = 70$ rats did not receive the treatment. On a critical test trial, the number of rats in each group turning left and right was as shown below. Use χ^2 to determine whether p_1 and p_2 differ significantly. (Do *not* make a correction for discontinuity.)

	Left	Right
Experimental	24	56
Control	36	34

13.5   The value of χ^2 for a 3×4 table is 15.0. Is this a significant value, if $\alpha = 0.05$?

13.6   It would have been possible to test the null hypothesis of Problem 13.1 by calculating Z. Do so. You should find that Z^2 is equal to the value of χ^2. Any value of χ^2 with 1 d.f. is always equal to Z^2.

13.7   In a discrimination experiment the probability of a correct response is $P = 1/3$. A subject is given $n = 18$ independent trials and makes $T = 10$ correct responses. (a) Use χ^2 to test the null hypothesis that we have a random sample from a binomial population with $P = 1/3$ and $Q = 2/3$. Do *not* make a correction for discontinuity. (b) Assume that T is normally distributed. Use the table of the standard normal curve to find the probability of $T \geq 10$. Do *not* make a correction for discontinuity. Is the value of Z^2 equal to the value of χ^2? (c) Would $\chi^2 = Z^2$ if a correction for discontinuity had been made in both tests?

13.8   For each of three independent groups of $n = 40$ rats, the number making a right turn and the number making a left turn at a choice point in a maze are given below. Test the null hypothesis that the three samples are from a common binomial population.

	Left	Right
Group 1	10	30
Group 2	30	10
Group 3	20	20

13.9   Two brands of a cola drink were presented to each of $n = 100$ students. Brand A was preferred by 65 students and Brand B by 35 students. (a) Use χ^2 to test the null hypothesis that the two brands are equally preferred. Make a correction for discontinuity. (b) Let $T = 65$ be a sum of $n = 100$ observations drawn from a binomial population. Use Z to test the null hypothesis that $P = Q = 1/2$. Make a correction for discontinuity. Is Z^2 equal to χ^2?

14

Measures of Association and Tests of Significance

14.1 Introduction

In this chapter, we describe three special cases of the correlation coefficient and also two additional measures of association that are applicable to certain kinds of data. In addition, we show how these various measures of association may be tested for significance.

14.2 The Rank-Order Coefficient

If we have two variables, X and Y, such that the observations for both variables consist of ranks, we can determine the degree of association between the two sets of ranks by means of the *rank-order coefficient of correlation*. The rank-order coefficient is simply a special case of correlation coefficient applicable to the case where both the X and Y variables consist of ranks. In this case

$$r = 1 - \frac{6\Sigma D^2}{n(n^2 - 1)} \tag{14.1}$$

where D = the difference between a pair of ranks
$\quad n$ = the number of ranks

We illustrate the calculation of r in Table 14.1, where we have $n = 10$ pairs of ranks. Column (4) gives the values of D and column (5) the values of D^2. The sum of column (5) gives ΣD^2 equal to 126. Then, substituting in (14.1), we have

$$r = 1 - \frac{(6)(126)}{10(10^2 - 1)} = 0.236$$

Table **14.1** — Ranks *X* and *Y* of 10 Observations

(1)	(2)	(3)	(4)	(5)
	Ranks			
Observation	*X*	*Y*	*D*	*D²*
a	1	6	−5	25
b	2	4	−2	4
c	3	5	−2	4
d	4	1	3	9
e	5	8	−3	9
f	6	10	−4	16
g	7	2	5	25
h	8	3	5	25
i	9	9	0	0
j	10	7	3	9
Σ	55	55	0	126

Formula (14.1) assumes that the observations for both variables are ranked without ties existing for any rank. However, if the number of tied observations is not large, and if the observations tied for a given rank are assigned the average of the ranks they would ordinarily occupy, the value of *r* defined by (14.1) will not differ too greatly from the value that would be obtained by taking the ties into consideration.

If *n* is equal to or greater than 10 then, to test the null hypothesis that the ranks of *X* and *Y* are independent, we calculate

$$t = \frac{r}{\sqrt{1-r^2}} \sqrt{n-2} \qquad (14.2)$$

with $n-2$ d.f. The probability associated with the *t* of (14.2) may be regarded as an approximation of the probability obtained from an exact test of the independence of the two sets of ranks.

For the data of Table 14.1, we have

$$t = \frac{0.236}{\sqrt{1-(0.236)^2}} \sqrt{10-2} = 0.69$$

with 8 d.f. If $\alpha = 0.05$, this is not a significant value, and we would not reject the null hypothesis that the *X* and *Y* ranks are independent.

14.3 The Phi Coefficient

We now consider a measure of association that is appropriate when *X* and *Y* both have only two classes of observations. We designate one of the *X* classes as 0 and the other as 1. We do the same for the *Y* classes. Then the possible paired (*X*,*Y*) values will be: (1,1), (1,0), (0,1), and (0,0). The

Table **14.2** — Schematic Representation of Paired X and Y Values with Two Classes for X and Two for Y

	X_0	X_1	Total
Y_1	a	b	$a + b$
Y_0	c	d	$c + d$
Total	$a + c$	$b + d$	n

schematic representation of the ordered pairs of X and Y values is shown in Table 14.2.

If we have a table with $r = 2$ rows and $c = 2$ columns, that is, a 2×2 table, such as Table 14.2, then we can use the *phi coefficient* as a measure of the degree of association between the row and column classifications. The phi coefficient is simply a special case of the correlation coefficient applicable to two variables when X can take only the values of $X = 1$ or $X = 0$ and Y also can take only the values of $Y = 1$ or $Y = 0$. In this case

$$r = \frac{bc - ad}{\sqrt{(a+c)(b+d)(a+b)(c+d)}} \tag{14.3}$$

where the letters refer to the cell and marginal frequencies of Table 14.2. We illustrate the calculation of r in terms of the data of Table 14.3. Substituting in (14.3), we have

$$r = \frac{(45)(80) - (45)(30)}{\sqrt{(125)(75)(90)(110)}} = 0.23$$

If we wish to determine whether the obtained value of $r = 0.23$ represents a significant degree of association between the two variables, we can test the general null hypothesis that the two variables are independent. The test statistic is

$$\chi^2 = nr^2 \tag{14.4}$$

with 1 d.f. For the data of Table 14.3, we have

$$\chi^2 = (200)(0.23)^2 = 10.58$$

By reference to the table of χ^2 with 1 d.f., we find that if $\alpha = 0.05$, the obtained value of χ^2 equal to 10.58 is significant. We thus reject the null

Table **14.3** — Distribution of Paired X and Y Values for 200 Observations

	X_0	X_1	Total
Y_1	45	45	90
Y_0	80	30	110
Total	125	75	200

hypothesis that the X and Y classifications are independent. The obtained value of r equal to 0.23 represents a small but significant degree of association between the two variables.

In calculating r, it is not necessary that the two classes of observations for the variables be ordered. The coefficient is applicable, in other words, to observations comprising nominal as well as ordered scales.

14.4 The Point-Biserial Coefficient

On some occasions, we may have one variable such that the values of the variable consist of only two classes comprising either a nominal or an ordered scale, whereas the values of the second variable may correspond to ordered measurements on an interval or ratio scale or simply ranks. We designate the variable with only two classes as the X variable and, for this variable, we have either $X = 1$ or $X = 0$. We represent the frequency of observations with $X = 0$ by $f_0 = n_0$ and the frequency of observations with $X = 1$ by $f_1 = n_1$. The second variable, for which we have an ordered set of measurements, we designate as the Y variable.

For the complete set of $n = n_0 + n_1$ observations, we then find the total sum of squares on the Y variable, or

$$\Sigma(Y - \bar{Y})^2 = \Sigma Y^2 - \frac{(\Sigma Y)^2}{n} \tag{14.5}$$

and (14.5) is precisely the same as the *total sum of squares* in the analysis of variance. Similarly, we can find the sum of squares between groups by calculating

$$\Sigma n_k (\bar{Y}_k - \bar{Y})^2 = \frac{(\Sigma Y_0)^2}{n_0} + \frac{(\Sigma Y_1)^2}{n_1} - \frac{(\Sigma Y)^2}{n} \tag{14.6}$$

where ΣY_0 is the sum of the Y values for the n_0 observations with $X = 0$ and ΣY_1 is the sum of the Y values for the n_1 observations with $X = 1$. We note that the sum of squares defined by (14.6) is precisely the same as the *treatment sum of squares* in an analysis of variance problem in which we have only two treatments or groups of observations.

As a measure of association between the X classification and the Y variable, we find

$$r = \sqrt{\frac{\Sigma n_k (\bar{Y}_k - \bar{Y})^2}{\Sigma (Y - \bar{Y})^2}} \tag{14.7}$$

and this coefficient is called the *point-biserial coefficient of correlation*. The point-biserial coefficient is also a special case of the correlation coefficient that is applicable when one variable X can take only a value of $X = 1$ and $X = 0$.

We illustrate the calculation of r for the data of Table 14.4. We have $\Sigma Y^2 = \Sigma Y_0^2 + \Sigma Y_1^2 = 486 + 473 = 959$ and $\Sigma Y = \Sigma Y_0 + \Sigma Y_1 = 58 + 77 =$

135. Then for the total sum of squares on the Y variable we have

$$\Sigma(Y - \bar{Y})^2 = 959 - \frac{(135)^2}{20} = 47.75$$

and for the sum of squares between groups we have

$$\Sigma n_k(\bar{Y}_k - \bar{Y})^2 = \frac{(58)^2}{7} + \frac{(77)^2}{13} - \frac{(135)^2}{20} = 25.40$$

Then, substituting in (14.7) with these two sums of squares, we have

$$r = \sqrt{\frac{25.40}{47.75}} = 0.73$$

Because $\Sigma n_k(\bar{Y}_k - \bar{Y})^2$ and $\Sigma(Y - \bar{Y})^2$ are always positive, the square root of (14.7) will also be positive. But the correlation coefficient, calculated in terms of any of the formulas given in Chapter 6, may in some cases be positive and in other cases negative. To make the sign of the point-biserial coefficient agree with the sign of the value of the correlation coefficient that would be obtained using other methods of calculation, we give the coefficient a positive sign if $\bar{Y}_1$ is larger than $\bar{Y}_0$; if $\bar{Y}_1$ is smaller than $\bar{Y}_0$, we give the coefficient a negative sign. In our example we have $\bar{Y}_0 = 58/7 = 8.29$ and $\bar{Y}_1 = 77/13 = 5.92$. Because $\bar{Y}_1$ is smaller than $\bar{Y}_0$, the correlation coefficient calculated in the usual way will be equal to -0.73.

To test the null hypothesis that the values of $\bar{Y}_0$ and $\bar{Y}_1$ are independent of the X classification, we first obtain the within group sum of squares. In the analysis of variance, this sum of squares corresponds to the *within treatment sum of squares* and can be obtained by subtraction. Thus

$$\text{Within} = \text{Total} - \text{Between groups} \qquad (14.8)$$

and for the data of Table 14.4, we have

$$\text{Within} = 47.75 - 25.40 = 22.35$$

and this sum of squares will have $n - 2$ d.f. The mean square within groups will, therefore, be equal to $22.35/18 = 1.24$. Because we have only two groups or classes of X, the sum of squares between groups will have only 1 d.f.

Taking F as the ratio of the mean square between groups and the mean square within groups, we have

$$F = \frac{25.40}{1.24} = 20.48$$

with 1 and 18 d.f. If $\alpha = 0.05$, then the obtained value of F equal to 20.48 would be regarded as significant, and the null hypothesis would be rejected. In essence, we conclude that the Y means for the observations in the class $X = 0$ and in the class $X = 1$ differ significantly. Thus the

Table 14.4 — Values of Y and Y^2 for Each of Two Classes of an X Variable

	X_0		X_1	
	Y	Y^2	Y	Y^2
	8	64	6	36
	8	64	6	36
	7	49	6	36
	8	64	7	49
	8	64	6	36
	9	81	6	36
	10	100	6	36
			6	36
			7	49
			5	25
			5	25
			3	9
			8	64
Σ	58	486	77	473

value of the mean on the Y variable is not independent of the X classification.

14.5 The Correlation Ratio: η_{YX}

Let us suppose that we have a variable X such that the classes of X form either a nominal or an ordered scale, and another variable Y such that the values of Y are at least ordered. For each of the k classes of X, we have $n_0, n_1, n_2, \ldots, n_k$ observations with $n = n_0 + n_1 + n_2 + \cdots + n_k$. It would be possible to find the mean Y value for each of the k classes of X. We wish to determine whether these mean Y values are in any way associated with the X classification. It is not necessary that the Y means be linearly related to the X classes and it is not necessary that the X classes be ordered, as we have already pointed out.

A measure of association between the Y means and the X classification is the *correlation ratio* or η_{YX}. We define this coefficient as

$$\eta_{YX} = \sqrt{\frac{\Sigma n_k (\bar{Y}_k - \bar{Y})^2}{\Sigma (Y - \bar{Y})^2}} \qquad (14.9)$$

where $\Sigma n_k (\bar{Y}_k - \bar{Y})^2$ is the sum of squares between groups and $\Sigma (Y - \bar{Y})^2$ is the total sum of squares, as described previously in the discussion of the analysis of variance. When we have only two X classes so that $k = 2$, then the correlation ratio is identical with the point-biserial coefficient of correlation.

The calculation of the correlation ratio is illustrated for the data of Table 14.5. We have $\Sigma Y^2 = 70 + 372 + 631 + 584 = 1657$ and $\Sigma Y = 18 + 52 + 61 + 48 = 179$. Then, for the total sum of squares, we have

$$\Sigma(Y - \bar{Y})^2 = 1657 - \frac{(179)^2}{23} = 263.9$$

and for the sum of squares between groups we have

$$\Sigma n_k (\bar{Y}_k - \bar{Y})^2 = \frac{(18)^2}{5} + \frac{(52)^2}{8} + \frac{(61)^2}{6} + \frac{(48)^2}{4} - \frac{(179)^2}{23} = 205.9$$

Then the value of the correlation ratio will be

$$\eta_{YX} = \sqrt{\frac{205.9}{263.9}} = 0.88$$

As a test of the null hypothesis that the Y means are independent of the X classification, we calculate

$$F = \frac{MS_T}{MS_W} \tag{14.10}$$

We note that the numerator of the F ratio is the mean square between groups and the denominator is the mean square within groups. The F ratio will therefore have $k - 1$ d.f. for the numerator and $n - k$ d.f. for the denominator.

For the data of Table 14.5, the sum of squares within groups is

$$\text{Within} = 263.9 - 205.9 = 58.0$$

with $n - k = 23 - 4 = 19$ d.f. Then

$$MS_W = \frac{58.0}{23 - 4} = 3.05$$

Table **14.5** — Values of Y for Each of Four Classes of an X Variable

	X_0	X_1	X_2	X_3
	5	10	12	14
	4	8	11	12
	4	8	11	12
	3	7	10	10
	2	6	9	
		5	8	
		5		
		3		
n	5	8	6	4
ΣY	18	52	61	48
ΣY^2	70	372	631	584

We have $k - 1 = 4 - 1 = 3$ d.f. for the sum of squares between groups and

$$MS_T = \frac{205.9}{4 - 1} = 68.63$$

Then, substituting in (14.10), we obtain

$$F = \frac{68.63}{3.05} = 22.5$$

with 3 and 19 d.f. With $\alpha = 0.05$, the value of F we have obtained is significant and the null hypothesis would be rejected. We conclude that the values of the Y means are not independent of the X classification.

14.6 The Contingency Coefficient: C

Suppose that we have two variables, X and Y, such that the classes of both variables constitute nominal scales. We wish to determine whether the row (Y) classification and the column (X) classification are independent. Suppose, for example, that $n = 250$ students have been asked to respond "Agree," "Undecided," or "Disagree" to two different statements in an opinion poll. We let response to one of the statements be the X variable and response to the other statement be the Y variable. Then we have the following possible ordered pairs of values of the two variables: (A_1, A_2), (A_1, U_2), (A_1, D_2), (U_1, A_2), (U_1, U_2), (U_1, D_2), (D_1, A_2), (D_1, U_2), (D_1, D_2), where the subscript 1 refers to response to the first statement and the subscript 2 refers to response to the second statement. Table 14.6 gives the frequency f with which each of these paired values occurred in the sample of $n = 250$. We wish to determine whether responses to the two statements are independent.

We have $10 + 35 + 50 = 95$ "Agree" responses to Statement 1 out of a total of $n = 250$ responses. Let us assume that the probability of an "Agree" response to Statement 1 is therefore

$$P(A_1) = \frac{95}{250} = 0.38$$

Similarly, we have $10 + 20 + 35 = 65$ "Agree" responses to Statement 2 and we assume that the probability of an "Agree" response to

Table **14.6** — Distribution of Paired Responses to Two Opinion Statements

	A_2	U_2	D_2	Total
A_1	10	35	50	95
U_1	20	40	10	70
D_1	35	40	10	85
Total	65	115	70	250

Statement 2 is

$$P(A_2) = \frac{65}{250} = 0.26$$

Then, if an "Agree" response to Statement 1 and an "Agree" response to Statement 2 are independent, we have as the probability of an "Agree" response to both statements

$$P(A_1 \cap A_2) = P(A_1)P(A_2)$$

and as the expected or theoretical frequency F of "Agree" responses to both statements, we have

$$nP(A_1 \cap A_2) = nP(A_1)P(A_2) \qquad (14.11)$$

In our example, we have $n = 250$ and $P(A_1) = 95/250$ and $P(A_2) = 65/250$, so that

$$nP(A_1 \cap A_2) = 250 \left(\frac{95}{250}\right)\left(\frac{65}{250}\right) = 24.7$$

In the same manner, we can find the other theoretical or expected frequencies F corresponding to each of the observed frequencies f in Table 14.6, under the assumption that the row and column classifications are independent. These are given in Table 14.7. It will be noted that the methods used in finding the theoretical frequencies F are exactly the same as those we described in the chapter on the χ^2 test of significance. Thus, we also have

$$\chi^2 = \sum \frac{(f-F)^2}{F} \qquad (14.12)$$

Substituting in (14.12) with the values of f and the corresponding values of F, as given in Table 14.7, we have

$$\chi^2 = \frac{(10-24.7)^2}{24.7} + \frac{(35-43.7)^2}{43.7} + \cdots + \frac{(10-23.8)^2}{23.8} = 53.38$$

Table **14.7** — Observed and Theoretical Frequencies of Paired Responses to Two Opinion Statements

Pair	f	F
A_1, A_2	10	24.7
A_1, U_2	35	43.7
A_1, D_2	50	26.6
U_1, A_2	20	18.2
U_1, U_2	40	32.2
U_1, D_2	10	19.6
D_1, A_2	35	22.1
D_1, U_2	40	39.1
D_1, D_2	10	23.8
Σ	250	250.0

The value of χ^2 obtained for the $r \times c$ table provides a test of significance of the null hypothesis that the row and column classifications are independent. The table of χ^2 is entered with $(r-1)(c-1)$ d.f. In our example, we have $(3-1)(3-1) = 4$ d.f. and according to the table of χ^2, a value of 13.277 or larger has a probability of 0.01 for 4 d.f., if the null hypothesis is true. If we have chosen $\alpha = 0.01$ then the obtained value of $\chi^2 = 53.38$ is significant and the null hypothesis would be rejected.

A measure of the degree of nonindependence or association between the row and column classifications is given by the *contingency coefficient*. This coefficient is defined as

$$C = \sqrt{\frac{\chi^2}{n+\chi^2}}$$ (14.13)

and for the example under consideration, we have

$$C = \sqrt{\frac{53.38}{250+53.38}} = 0.42$$

14.7 Significance of the Correlation Coefficient

If we have calculated the correlation coefficient r, as described in Chapter 6, we may wish to test the null hypothesis that the two variables are not linearly related. As a test of significance we have, using (14.2),

$$t = \frac{r}{\sqrt{1-r^2}} \sqrt{n-2}$$

with $n-2$ d.f.

Assume, for example, that the correlation coefficient between scores on an arithmetic test and a reading achievement test is found to be $r = 0.53$ for a sample of $n = 32$ paired observations. Substituting in the above expression, we have

$$t = \frac{0.53}{\sqrt{1-(0.53)^2}} \sqrt{32-2} = 3.4$$

with $32-2$ d.f. Entering the table of t in the Appendix with 30 d.f., we find that $t = 3.4$ is a significant value with a probability of less than 0.01, if the null hypothesis is true. We conclude, therefore, that there is a significant linear relationship between scores on these two tests.

REVIEW OF TERMS

between group sum of squares
contingency coefficient
correlation ratio

point-biserial coefficient
rank-order coefficient
ranks

nominal scale

total sum of squares

ordered variable

unordered variable

phi coefficient

within group sum of squares

REVIEW OF SYMBOLS

r

f

χ^2

$\Sigma (Y - \bar{Y})^2$

F

$\Sigma n_k (\bar{Y}_k - \bar{Y})^2$

η_{YX}

n_0

t

n_1

C

QUESTIONS AND PROBLEMS

14.1   Out of 100 students, 60 gave the Right response to Item 1 and 45 gave the Right response to Item 2. Thirty-five students gave the Right response to both items. (a) Find the value of r. (b) Test the null hypothesis that responses to the two items are independent.

14.2   For a 3×5 table with a total of 110 observations, we have χ^2 equal to 15.0. (a) What is the value of the contingency coefficient? (b) Test the null hypothesis that the row and column classifications are independent.

14.3   There are 11 males and 15 females registered in a university course. The scores of the 11 males and of the 15 females on an attitude scale are given below.

Males	Females
5	5
3	6
1	3
4	4
3	2
2	7
3	3
3	3
3	1
3	6
4	3
	5
	4
	2
	3

(a) Find the value of r. (b) Test the null hypothesis that the mean scores on the test are independent of the sex classification.

14.4 We have the following values of Y for five categories of X.

X_1	X_2	X_3	X_4	X_5
2	2	5	7	3
4	4	5	6	2
4	4	4	4	4
2	6	5	8	4
3	3	3		2
3	7			3
	5			1
				3

(a) Find the value of the correlation ratio. (b) Test the null hypothesis that the values of the Y means are independent of the X classification.

14.5 Two student judges ranked 10 candidates for campus beauty queen for attractiveness. The ranks assigned each candidate by the two students are given below.

Candidate	A	B	C	D	E	F	G	H	I	J
Judge 1	1	2	3	4	5	6	7	8	9	10
Judge 2	1	2	5	6	4	3	9	10	7	8

(a) Find the value of the rank-order coefficient. (b) Test the null hypothesis that the ranks assigned by the two judges are independent.

14.6 A correlation coefficient equal to 0.32 was obtained between an X and Y variable for 27 paired measurements. Test the null hypothesis that the X and Y values are not linearly related.

14.7 Calculate the point-biserial coefficient for the data given below. Y_0 are the Y values associated with $X = 0$ and Y_1 are the Y values associated with $X = 1$.

Y_0	Y_1
2	5
6	3
4	1
10	2
8	4

14.8 We have a correlation coefficient equal to 0.60 for $n = 11$ paired measurements. Test the null hypothesis that the X and Y values are not linearly related.

14.9 Calculate the contingency coefficient for the following table in which one variable A has three categories and the other variable B has two categories:

	B_1	B_2
A_1	10	30
A_2	20	20
A_3	30	10

14.10 Calculate the phi coefficient for the data given below.

	X_0	X_1
Y_1	10	20
Y_0	30	40

14.11 Calculate the rank-order correlation coefficient for the two sets of ranks given below.

X	Y
1	2
2	1
3	3
4	5
5	4

14.12 For the ranks given in Problem 14.11, calculate the correlation coefficient using

$$r = \frac{\Sigma xy}{\sqrt{(\Sigma x^2)(\Sigma y^2)}}$$

Is the value of r equal to the value of r obtained in Problem 14.11?

14.13 Calculate the phi coefficient for the table given below.

	X_0	X_1
Y_1	4	6
Y_0	6	4

14.14 In Problem 14.13, we have ten values of $Y = 1$ and ten values of $Y = 0$. Similarly, we have ten values of $X = 1$ and ten values of $X = 0$. Calculate the correlation coefficient for the same problem, using

$$r = \frac{\Sigma xy}{\sqrt{(\Sigma x^2)(\Sigma y^2)}}$$

Is the value of r equal to the value of r obtained in Problem 14.13?

14.15 Find the square of the point-biserial coefficient for the data given below.

X_0	X_1
1	4
2	5
3	

14.16 For Problem 14.15 we have two values of $X = 1$ and three values of $X = 0$. For the same problem calculate the square of the correlation coefficient using

$$r^2 = \frac{(\Sigma xy)^2}{(\Sigma x^2)(\Sigma y^2)}$$

Is the value of r^2 equal to the value of r^2 obtained in Problem 14.15?

15

Random Errors of Measurement

15.1 Introduction

Every set of measurements is subject to errors of observation. If, for example, we had several hundred objects of varying lengths and we measured the length of each object twice, we would not expect all of the pairs of measurements to be precisely the same. Slight errors of observations are apt to be present, despite efforts to reduce these to a minimum. Sometimes the second measurement might be slightly less than the first, sometimes it might be slightly more, and in other cases we might have exactly the same recorded value for both observations.

We may distinguish between *systematic* and *random* errors of observation or measurement. Systematic errors are errors that tend to result in a consistent over- or underestimation of the true measurement. For example, suppose that we have several hundred objects whose true weights are known. These objects are now weighed on a scale, and the resulting scale values are recorded. If we now subtract the true value for each weight from the corresponding observed value, we describe the resulting discrepancy as an error of measurement.

If, in general, the errors of measurement tended to be consistently positive or consistently negative in sign, we would regard them as systematic errors. If, on the other hand, we found that the positive and negative errors occurred with approximately the same frequency and that the sum of the negative errors was approximately equal to the sum of the positive errors, we would regard them as random or chance errors. It should be clear that if the sum of the positive errors is equal to the sum of the negative errors, then the average error would be equal to zero. It is not unusual to assume that a distribution of random errors is approximately normal in form with the mean equal to zero.

A good research worker or experimenter makes every effort to free his measurements from systematic errors and to reduce his random errors to a minimum. One reason for this, as we shall see, is that although random errors do not influence the mean value of a set of observations, they do tend to increase the value of the variance and consequently the value of the standard error of the mean. Random errors of measurement can perhaps never be eliminated completely, and the notion of a true measurement, at least in the social and biological sciences at the present time, must remain a theoretical concept. We can, however, work with the notion of a true measurement even though we recognize the difficulties involved in actually obtaining such a measurement.

Assume, for example, that we have given to a large group of subjects a test designed to measure some aspect of arithmetic ability. We have available the scores of each subject on the test. These scores do not necessarily correspond to the true scores of the subjects, and we make the assumption that the error of measurement, which will be the difference between the observed score and the true score, is random rather than systematic. *We shall hold to this assumption throughout the rest of the discussion in this chapter.*

15.2 Assumptions Regarding Random Errors

From the statements made above we may define a random error of measurement associated with the values of a given variable X as

$$e_X = X - X_T \tag{15.1}$$

where X is the observed value and X_T is the true value. Similarly, we define a random error of measurement associated with the values of another variable Y as

$$e_Y = Y - Y_T \tag{15.2}$$

where Y is the observed value and Y_T is the true value.

If we make a sufficiently large number of observations and if e_X and e_Y are random errors, then we shall assume that the average values of both e_X and e_Y are equal to zero. Furthermore, we shall assume that random errors are uncorrelated with true scores and also with each other. In other words, we assume that if we have a sufficiently large number of observations, then

$$r_{X_T e_X} = r_{Y_T e_Y} = r_{Y_T e_Y} = r_{X_T e_Y} = r_{e_X e_Y} = 0 \tag{15.3}$$

In an earlier chapter, we had as one of the formulas for the correlation coefficient between two variables

$$r_{XY} = \frac{\Sigma xy}{\sqrt{\Sigma x^2 \Sigma y^2}}$$

and if r_{XY} is equal to zero, then it must also be true that the product sum Σxy is also equal to zero, because $\Sigma xy = r_{XY}\sqrt{\Sigma x^2 \Sigma y^2}$. Thus, all of the product sums

$$\Sigma x_T e_X, \Sigma x_T e_Y, \Sigma y_T e_X, \Sigma y_T e_Y, \text{and} \Sigma e_X e_Y$$

for the correlation coefficients given by (15.3) must also be equal to zero.

15.3 Influence of Random Errors on the Mean

From (15.1) it follows that an observed value of X will be given by

$$X = X_T + e_X \qquad (15.4)$$

If we sum both sides of (15.4) and divide by the number of observations, then we have

$$\bar{X} = \bar{X}_T \qquad (15.5)$$

because we have assumed that the mean of a distribution of random errors will be equal to zero. Consequently, we may conclude that the value of the mean of a large number of observations will not be influenced or biased by the presence of random errors.[1]

15.4 Influence of Random Errors on the Variance

If the mean of a distribution of random errors is equal to zero, then e_X is already in deviation form. And, because $\bar{X}$ is equal to $\bar{X}_T$, we may write (15.4) in deviation form to obtain

$$X - \bar{X} = X_T - \bar{X}_T + e_X$$

or

$$x = x_T + e_X \qquad (15.6)$$

If we now square both sides of (15.6) and sum over all observations, we obtain

$$\Sigma x^2 = \Sigma x_T^2 + \Sigma e_X^2 + 2\Sigma x_T e_X \qquad (15.7)$$

According to (15.3), we have $r_{x_T e_X}$ equal to zero and therefore the product sum $\Sigma x_T e_X$ must be equal to zero. Thus we have

$$\Sigma x^2 = \Sigma x_T^2 + \Sigma e_X^2 \qquad (15.8)$$

and we note that the observed sum of squares Σx^2 must be greater than the true sum of squares Σx_T^2, if random errors of measurement are present. Because the observed sum of squares, divided by $n-1$, gives the

---

[1]This, of course, may not be true for any particular sample set of measurements in which the sum of the positive errors may differ slightly from the sum of the negative errors. We are dealing here, however, with theoretical notions rather than with actualities and, in theory, Σe will approach zero as n becomes indefinitely large.

variance, we must conclude that the observed variance will be greater than the true variance, if random errors of measurement are present. Furthermore, because the standard error of a mean is given by $s_{\bar{X}} = s_X/\sqrt{n}$, random errors of measurement will serve to increase the standard error of the mean over the value that would be obtained with measurements free from random errors.

15.5 Influence of Random Errors on the Correlation Coefficient

Let us now see what influence random errors of measurement will have on the correlation between two variables. Expressing both the X and Y values in the form given by (15.6), the numerator of the correlation coefficient will be given by

$$\Sigma xy = \Sigma (x_T + e_X)(y_T + e_Y) \tag{15.9}$$

Expanding the right side of (15.9) and then summing, we obtain

$$\Sigma xy = \Sigma x_T y_T + \Sigma x_T e_Y + \Sigma y_T e_X + \Sigma e_X e_Y \tag{15.10}$$

But the various product sums in (15.10) will involve the corresponding correlation coefficients and we have assumed that random errors will be uncorrelated with true scores and with each other. Therefore,

$$\Sigma x_T e_Y = \Sigma y_T e_X = \Sigma e_X e_Y = 0 \tag{15.11}$$

and thus we have

$$\Sigma xy = \Sigma x_T y_T \tag{15.12}$$

We see that the product sum Σxy will be equal to the true product sum $\Sigma x_T y_T$, despite the presence of random errors. But the denominator of the correlation coefficient will involve the sum of squares for X and also the sum of squares for Y. Thus

$$r_{XY} = \frac{\Sigma xy}{\sqrt{\Sigma x^2 \Sigma y^2}}$$

We have just shown that $\Sigma xy = \Sigma x_T y_T$. Formula (15.8) provides an identity for Σx^2 and we have a similar expression for Σy^2. Substituting these expressions in the formula for the correlation coefficient, we obtain

$$r_{XY} = \frac{\Sigma xy}{\sqrt{(\Sigma x_T^2 + \Sigma e_X^2)(\Sigma y_T^2 + \Sigma e_Y^2)}} \tag{15.13}$$

It is clear from (15.13) that if random errors of measurement are present in X and Y, the denominator of the correlation coefficient will be larger than would be the case for measurements free from such errors. We may conclude, therefore, that random errors of measurement will tend to reduce the value of the observed correlation coefficient in comparison with the value that would be obtained in the absence of such errors.

15.6 Reliability Coefficient: Comparable Forms

Suppose that we have available two forms of the same psychological or educational test. We shall assume that if we gave both forms of the test to a large group of subjects we would find approximately the same means and variances for the two sets of scores. The two scores for each subject, however, will not necessarily be identical, if the scores involve random errors of measurement. The correlation coefficient between the two sets of scores will take the form

$$r_{X_1X_2} = \frac{\Sigma x_1 x_2}{\sqrt{\Sigma x_1{}^2 \Sigma x_2{}^2}} \tag{15.14}$$

where $r_{X_1X_2}$ = the correlation coefficient between the two forms of a test
    x_1 = a deviation score on one form of the test
    x_2 = a deviation score on the second form of the test

The product sum in the numerator of the correlation coefficient will be similar to that given by (15.9). Substituting this expression for $\Sigma x_1 x_2$, we obtain

$$r_{X_1X_2} = \frac{\Sigma(x_T + e_1)(x_T + e_2)}{\sqrt{\Sigma x_1{}^2 \Sigma x_2{}^2}} \tag{15.15}$$

where e_1 and e_2 are random errors associated with the scores on the first and second forms of the test, respectively.

Expanding the numerator of the right-hand side of the above expression and summing, we have

$$r_{X_1X_2} = \frac{\Sigma x_T{}^2 + \Sigma x_T e_2 + \Sigma x_T e_1 + \Sigma e_1 e_2}{\sqrt{\Sigma x_1{}^2 \Sigma x_2{}^2}} \tag{15.16}$$

Because we have assumed that the random errors will be uncorrelated with true scores and with each other, $\Sigma x_T e_2$, $\Sigma x_T e_1$, and $\Sigma e_1 e_2$ will be equal to zero. Therefore

$$r_{X_1X_2} = \frac{\Sigma x_T{}^2}{\sqrt{\Sigma x_1{}^2 \Sigma x_2{}^2}} \tag{15.17}$$

Under the assumption of equal variances, it follows that $\Sigma x_1{}^2$ and $\Sigma x_2{}^2$ will be equal, and we may drop the subscripts and take the square root of the denominator to obtain

$$r_{X_1X_2} = \frac{\Sigma x_T{}^2}{\Sigma x^2} \tag{15.18}$$

From (15.8) we note that $\Sigma x_T{}^2 = \Sigma x^2 - \Sigma e_X{}^2$. With the assumption that $\Sigma e_1{}^2 = \Sigma e_2{}^2 = \Sigma e^2$, we have

$$r_{X_1X_2} = \frac{\Sigma x^2 - \Sigma e^2}{\Sigma x^2}$$

$$= 1 - \frac{\Sigma e^2}{\Sigma x^2} \tag{15.19}$$

Dividing both Σe^2 and Σx^2 by $n-1$, we obtain the following expression:

$$r_{X_1X_2} = 1 - \frac{s_e^2}{s_X^2} \qquad\qquad 15.20)$$

The correlation coefficient given by (15.20) is called a *reliability coefficient*. The ratio s_e^2/s_x^2 is the ratio between the error variance and the observed variance. It is apparent from (15.20) that, if the error variance is as great as the observed variance, the reliability coefficient will be equal to zero. On the other hand, the smaller the error variance, in comparison with the observed variance, the larger the reliability coefficient. In the limiting case, with no random errors of measurement, so that s_e^2 is equal to zero, the reliability coefficient will be equal to 1.00.

15.7 Test-Retest Reliability Coefficient

If a psychological or educational test were developed so that two comparable forms of the test were available, the reliability coefficient could be obtained by testing a large group of subjects with both forms and correlating the resulting paired scores. We assume, of course, that the two forms could be made comparable by the careful selection of items used in each form.

The construction of the single form of a test involves a great deal of work, and the construction of two comparable forms more than doubles the labor involved in constructing a single form. In many instances, therefore, a test is available in only one form. When this is the case, the reliability coefficient for scores on the test is often obtained by administering the same subjects the single form of the test on two different occasions and then correlating the paired scores for the subjects.

Certain difficulties are involved in this procedure in that if the interval separating the two administrations of the test is quite short, memory, practice effects, and other factors may influence the scores obtained from the second administration. If the time interval is quite long and if the variable that the test is designed to measure is such that subjects may be expected to change so that they may have different scores on the second administration, the correlation coefficient will be influenced by these changes. The same difficulties would be involved in administering two comparable forms of the test, except that, if we have different items in the two forms, we might rule out memory as a contributing influence.

15.8 Split-Half Reliability Coefficients

Another method of estimating the reliability of scores on a test when only one form of the test is available is to divide the items on the test in such a way as to yield two scores. If, for example, a score is obtained from the odd-numbered items and another score is obtained from the even-numbered items, these two scores may be correlated. The resulting correlation coefficient is called a *split-half reliability coefficient.*

It is shown in textbooks dealing with the theory of test construction that the reliability coefficient of a test is influenced by the number of items in the test. In general, the larger the number of items, with other things being equal, the larger the value of the reliability coefficient. The split-half reliability coefficient gives the correlation between the two halves of a test and consequently refers to the reliability of a test with one-half the number of items that the test itself contains. What we desire to know is not the reliability of the scores obtained from the split-halves, but rather the reliability of the test in its original length, that is, of scores based on all of the items.

The reliability of the scores on the total test can be estimated from the split-half reliability coefficient. For example, if we have a reliability coefficient based on the correlation between two sets of n items, and we wish to estimate the reliability coefficient for a test based on the correlation between two sets of k items, then

$$r_{kk} = \frac{mr_{nn}}{1 + (m-1)r_{nn}} \qquad (15.21)$$

where r_{kk} = the estimated reliability of the test with k items
$m = k/n$
r_{nn} = the reliability coefficient of scores based on n items each

Now, if we have divided a test into two halves, so that each half contains n items, the scores on the total test will be based on $2n$ items. Then $m = k/n = 2n/n = 2$. Thus, for the estimated reliability of the complete test, we have

$$r_{kk} = \frac{2r_{nn}}{1 + r_{nn}} \qquad (15.22)$$

15.9 The K-R 20 Measure of Internal Consistency

If a test has n items and if we divide the items into those that are odd-numbered and those that are even-numbered, this subdivision of the test items into halves is only one of the

$$\frac{n!}{(n/2)!(n/2)!}$$

possible ways in which the n items may be divided into two equal sets. There is no necessary intrinsic merit in any one of the possible ways in which the items can be divided into two equal sets, and it seems reasonable that the values of the split-half reliability coefficients obtained from these various splits would not all be the same. Which one of these values should be regarded as the split-half reliability coefficient?

Another method of estimating the reliability coefficient when only the scores on a single test are available is to calculate the Kuder-Richardson Formula 20 (K-R 20) coefficient. The K-R 20 coefficient gives a unique value that is not dependent upon the particular split of the items into two halves. The K-R 20 coefficient is defined by

$$r_{X_1X_2} = \left(\frac{n}{n-1}\right)\frac{s_X^2 - \Sigma p_i q_i}{s_X^2} \qquad (15.23)$$

where n = the number of items in the test
s_X^2 = the variance of the scores on the test
$\Sigma p_i q_i$ = the sum of the item variances

In an earlier chapter, it was pointed out that the variance of a sum of n independent variables is equal to the sum of the variances of the variables. We know that if the items in a test are scored either $X = 1$ or $X = 0$, then the item variances will be equal to $p_i q_i$, where p_i is the proportion of subjects giving the response $X = 1$ and q_i is $1 - p_i$ or the proportion of subjects giving the response $X = 0$. Then, if response to each of the items in a test is independent or uncorrelated with the response to every other item, the variance of the total scores on the test would be

$$s_X^2 = \Sigma p_i q_i \qquad (15.24)$$

or, in other words, the sum of the item variances.

If response to the items in a test are not independent, then the variance of the total scores on the test will be

$$s_X^2 = \Sigma p_i q_i + \Sigma r_{ij}\sqrt{p_i q_i}\sqrt{p_j q_j} \qquad (15.25)$$

where r_{ij} is the correlation between response to Item i and Item j, $\sqrt{p_i q_i}$ is the standard deviation of the responses to Item i, and $\sqrt{p_j q_j}$ is the standard deviation of the responses to Item j. Each of the $n(n-1)$ possible values of $r_{ij}\sqrt{p_i q_i}\sqrt{p_j q_j}$ is a covariance and it is obvious that if response to each item is uncorrelated with response to every other item, so that for all of the $n(n-1)$ covariance terms we have $r_{ij} = 0$, then we would have $s_X^2 = \Sigma p_i q_i$ and the K-R 20 value of the test would be equal to zero.

In essence, then, the K-R 20 value of a test indicates the degree to which the responses to the items in a test are positively intercorrelated. If item responses are positively intercorrelated, then we have some assurance that the items in the test are measuring consistently whatever it is that they are measuring. For this reason, the K-R 20 value of a test is sometimes referred to as a measure of the *internal consistency* of a test.

In addition to knowing the number of items and the mean and standard deviation of the scores on a test, the K-R 20 value is one of the most important properties of a test. If a test has a high K-R 20 value, then it will also tend to have a high split-half reliability coefficient, regardless of the particular manner in which the items are subdivided into two equal sets. In general, it will also be true that a test that has a high K-R 20 value will also tend to have a high test-retest coefficient, at least over short intervals of time.

On the other hand, it does not follow that simply because a test has a high test-retest coefficient, the K-R 20 value of the test will also be high. It is also possible for the correlation between two comparable forms of a test to be high without the necessity of either form having a very high K-R 20 value.

15.10 Correction for Attenuation

It is sometimes of interest to ask what the maximum correlation would be between two variables X and Y, if no random errors were present in either set of measurements. This correlation would be of the form

$$r_{X_T Y_T} = \frac{\Sigma x_T y_T}{\sqrt{\Sigma x_T^2 \Sigma y_T^2}} \tag{15.26}$$

We have already shown that $\Sigma xy = \Sigma x_T y_T$. We may also observe that if we multiply both sides of (15.18) by Σx^2, we obtain

$$r_{X_1 X_2} \Sigma x^2 = \Sigma x_T^2 \tag{15.27}$$

and we would have a similar expression, $r_{Y_1 Y_2} \Sigma y^2 = \Sigma y_T^2$, for the true sum of squares for Y. Substituting these identities in (15.26) we have

$$r_{X_T Y_T} = \frac{\Sigma xy}{\sqrt{(r_{X_1 X_2} \Sigma x^2)(r_{Y_1 Y_2} \Sigma y^2)}} \tag{15.28}$$

Rearranging these terms, we obtain

$$r_{X_T Y_T} = \frac{\Sigma xy}{\sqrt{\Sigma x^2 \Sigma y^2}} \frac{1}{\sqrt{r_{X_1 X_2} r_{Y_1 Y_2}}}$$

in which the first expression on the right is the observed coefficient of correlation between X and Y. Therefore

$$r_{X_T Y_T} = \frac{r_{XY}}{\sqrt{r_{X_1 X_2} r_{Y_1 Y_2}}} \tag{15.29}$$

The coefficient defined by (15.29) is called the correlation coefficient *corrected for attenuation*. The correction is of theoretical interest in that it provides an estimate of the correlation between X and Y if we had true measures of the two variables, free from random errors of measurement.

If we multiply both sides of (15.29) by the denominator of the right-hand side, we obtain

$$r_{XY} = r_{X_T Y_T} \sqrt{r_{X_1 X_2} r_{Y_1 Y_2}} \qquad (15.30)$$

We may observe from (15.30) that if we had perfectly reliable measures of X and Y, so that the resulting reliability coefficients would be equal to 1.00, the observed correlation coefficient would be equal to the true correlation. If one of the two variables is perfectly reliable and the correlation between the true scores is equal to 1.00, then the observed correlation coefficient cannot be greater than the square root of the reliability coefficient of the second variable. If the reliability coefficient of the X variable is equal to the reliability coefficient of the Y variable, then the observed correlation coefficient cannot be greater than the common reliability coefficient because $r_{X_T Y_T}$ cannot be greater than 1.00. Finally, if the correlation between the true scores is equal to 1.00, but if random errors of measurement are present in both variables, the observed correlation coefficient cannot exceed the square root of the product of the two reliability coefficients.

15.11 Validity Coefficients

The correlation between scores on a Test X and some independent criterion that it is hoped the scores will predict, is called a *validity coefficient*. For example, there are many tests that have been designed to measure the academic success of students in college, the criterion being regarded as the grade-point average of the students. If we correlate scores on the test with grade-point averages, the resulting correlation coefficient is a *validity* coefficient. The higher the validity coefficient, the better we can predict grade-point averages in terms of the regression equation. If the test were perfectly valid, that is, if the correlation were equal to 1.00, it would be possible to predict precisely the grade-point average of each student from his score on the test. If the test fails to correlate with grade-point average, we say that the test is not valid for this purpose.

Any particular psychological test may, of course, have many different validity coefficients in terms of the degree to which it correlates with different criteria, and it is nonsense to talk about validity in the abstract. The validity of a test is always with reference to some particular criterion and not criteria in general. A test may have a specified degree of validity for predicting academic success and no validity whatsoever for predicting income or popularity or some other criterion.

Validity, as measured by the correlation between a test X and some criterion Y, is closely related to the reliability coefficient of the test $r_{X_1 X_2}$ and also to the reliability coefficient of the criterion $r_{Y_1 Y_2}$. This follows from

(15.30), where

$$r_{XY} = r_{X_T Y_T} \sqrt{r_{X_1 X_2} r_{Y_1 Y_2}}$$

It is obvious, for example, that we can have maximum validity in terms of r_{XY} only when we have perfect reliability in terms of $r_{X_1 X_2}$ and $r_{Y_1 Y_2}$. If the reliability coefficients of the test and criterion are equal, but less than 1.00, then the observed validity coefficient cannot be greater than the common reliability coefficient, because the true validity coefficient cannot be greater than 1.00. If the test was perfectly reliable and if the true validity coefficient $r_{X_T Y_T}$ was also equal to 1.00, the observed validity coefficient r_{XY} could not be greater than the square root of the reliability coefficient of the criterion.

REVIEW OF TERMS

attenuation correction	split-half reliability coefficient
internal consistency	systematic errors
random errors	true score
reliability coefficient	validity coefficient

REVIEW OF SYMBOLS

e_X	x_T	Σx_T^2
X_T	x	Σx^2
$r_{X_T Y_T}$	s_e^2	r_{XY}
$r_{X_1 X_2}$	K-R 20	$p_i q_i$

QUESTIONS AND PROBLEMS

15.1   If the split-half reliability coefficient of a test of 60 items is 0.80, then what is the estimated reliability of the complete test?

15.2   If $r_{X_1 X_2} = 0.81$, $r_{Y_1 Y_2} = 0.64$, and $r_{X_T Y_T} = 1.00$, then what is the estimated maximum value of the observed correlation between X and Y?

15.3   If $r_{X_1 X_2} = 0.81$, $r_{X_T Y_T} = 1.00$, and $r_{XY} = 0.90$, then what is the value of $r_{Y_1 Y_2}$?

15.4   If the split-half reliability of a test of 20 items is 0.60, then what is the estimated reliability of the complete test?

15.5   If the split-half reliability of a test of 20 items is 0.60, then how many items would the test have to have in order to obtain an estimated reliability coefficient of 0.90?

15.6   Under the assumptions stated in the chapter, what influence will random errors of measurement have on each of the following statistics:
(a) the mean
(b) the standard deviation
(c) the covariance
(d) the correlation coefficient
(e) the standard error of a mean

15.7   A test consists of $n = 18$ items. The proportion of the total number of subjects giving the keyed response to each of the 18 items is shown below. The standard deviation on the test for a sample of 229 males is 4.09. (a) What is the mean score on the test? (b) What is the K-R 20 value of the test?

Item	p_i
1	0.72
2	0.86
3	0.90
4	0.91
5	0.93
6	0.95
7	0.51
8	0.93
9	0.82
10	0.87
11	0.57
12	0.57
13	0.61
14	0.78
15	0.53
16	0.70
17	0.63
18	0.82

15.8   A test consists of $n = 5$ items. The proportion of the total number of subjects giving the correct response to each of the five items is shown below. The standard deviation of the test is 2.13. Determine the Kuder-Richardson coefficient for the test.

Item	p_i
1	0.5
2	0.6
3	0.7
4	0.6
5	0.5

15.9   For a group of subjects we have paired (X, Y) measurements on two variables. The X measurements are scores on a test and in scoring the test a systematic error was made which resulted in increasing each value of X by two points. Explain what influence, if any, this systematic constant error will have on each of the statistics listed below, compared with the value that would have been obtained if the error had not been made.
(a) the mean of the X values
(b) the standard deviation of the X values
(c) the covariance of the X and Y values
(d) the correlation coefficient between the X and Y values
(e) the standard error of the mean of the X values

16

Samples in Research

16.1 Introduction

When, as may often be the case, we are interested in making inferences on the basis of statistics derived from samples about a characteristic of a population or universe, certain assumptions are necessary. We must assume, for example, that the sample is representative of the population of interest. If the population of observations of interest consists of the measured heights of all male students enrolled in a given university, the sample obtained by measuring only the heights of the members of the basketball squad would not be representative of the population specified. Unless we have information about the manner in which the observations are selected, we have no way of determining whether or not the sample is representative of the population in which we are interested.

16.2 Systematic and Random Errors

Samples used in experiments and research, and consequently statistics derived from these samples, are subject to the two kinds of error mentioned in the previous chapter: systematic errors and random errors. Both of these kinds of errors may enter into the selection of the observations and into the values of the observations. Systematic errors, as was pointed out in the previous chapter, are those that result in biasing observations in one direction. The standard error formulas and tests of significance we have described do not provide any estimate of the direction or magnitude of systematic errors in the selection of observations. Systematic errors of this kind can possibly be detected by a logical examination of the manner in which the sample was selected.

The standard error formulas do provide estimates of the errors that can be attributed to random sampling from a specified population. The standard error of the mean, for example, estimates the variation to be expected in sample means of size n in random sampling from a specified population. It is randomness that provides a frame of reference for interpreting the results obtained from standard error formulas.

We have stated that random and systematic errors may also enter into the values of observations. As an example of a systematic error, assume that an observer, in reading the pointer of a dial, tends to read the value as higher than it actually is. If he does this consistently, the mean of a series of such observations will be greater than if the location of the pointer had been read accurately. A systematic error of the kind described will result in a bias in the sample mean. If the systematic error is a *constant* error (that is, if it is always of the same magnitude) then it will not result in a bias in the standard error calculated from a sample, because the variance of a set of observations is unchanged by the addition of a constant to each value of a variable.

If errors of measurement are random, in the sense described in the previous chapter, then the presence of such errors will, in general, tend to increase the variance of the observations and thus to increase the standard error of the mean. Standard error formulas are thus sensitive to random errors of measurement as well as to random errors of selection. Because it is to our advantage in research to have standard errors as small as possible, anything that will reduce random errors of measurement is desirable. The psychometrician refers to a psychological or educational test that is relatively free from random errors of measurement as a *reliable* test. In measurements of all kinds, not just tests, we desire reliable measuring instruments.

16.3 Normal Distribution and Normal-Distribution Free Tests

We have emphasized in the discussion of tests of significance the notion of random selection from a specified population. We now consider the nature of the specified population in somewhat greater detail. We said, for example, in the discussion of the t and F tests, that we assumed that samples were randomly selected from identical normal populations. If we are primarily interested in the mean of a sample, we know that the random sampling distribution of the mean tends toward normality as the sample size n increases. We provided a number of illustrative sampling distributions to indicate that this was the case, even when the population was rectangularly distributed. We found, for example, that the random sampling distribution of p, the mean of a sample from a binomial population, was, in fact, approximately normally distributed when both nP and nQ were equal to or greater than 5. The assumption of normality of distribution of the individual observations in the population is one that can be greatly relaxed without seriously distorting conclusions based on tests of significance concerning means, as long as random selection of observations is involved and provided n, the sample size, is reasonably large.

There are some tests of significance that can be used to test the null hypothesis that samples have been randomly drawn from identical populations, that is, that do not involve the notion of normality of distribution. These tests are often described as *normal-distribution free* tests.[1] The assumptions involved in these tests, as applied to differences in means, assume random selection and are, in other respects, similar to those for the t test; that is, they still assume random selection and equivalence of population variances. The only thing they do not assume is that the populations are normally distributed. In discussions of these tests, it is often pointed out that they are not very sensitive to differences in population variances. We emphasize that this is also true of the t test for the difference between two means and of the F test for the differences among several means, provided that each mean is based on the same number of observations. There is ample evidence to indicate that normal distribution tests are, in fact, relatively insensitive to differences in population variances in their application to problems of the kind described in this text.

With respect to both normal distribution and normal-distribution free tests, all assumptions (other than that of random selection) can thus be greatly relaxed without voiding conclusions based on the test.

16.4 Finite Populations

The populations that we randomly sample in research and experimentation are almost never, if ever, infinite, although they are often assumed to be so. They are, in general, finite or limited. Thus, the standard error formulas should, in general, have a correction factor for sampling from a finite population. We have already indicated, however, that the correction factor is relatively unimportant if the sampling fraction n/N is less than 20 percent. Because the correction factor tends to reduce the estimated standard error of the mean from that obtained without the correction factor, failure to use it may mean that if we have chosen $\alpha = 0.05$, the true significance level is somewhat smaller. A test of significance that gives a significant result without the correction factor will, in other words, also give a significant result if we had used the correction factor. In suggesting

---

[1] They are also referred to as nonparametric tests.

that the correction factor may be ignored, we are merely following fairly standard practice in research, with the understanding that we thereby decrease the power of a test.

A *finite population* is one in which we could, in theory, denote each of the observations in the population. In psychology, education, and the behavioral sciences, we most often think of observations as being associated with organisms. In this instance, a random sample of observations would be obtained by randomly selecting organisms from the population specified. If this population is finite, then the individuals can be enumerated, and samples can be randomly selected by means of a table of random numbers. If this is done, then a conclusion based on an experiment in which a test of significance is applied to the sample observations can be assumed to hold for the population enumerated. If the enumerated population can be said to be representative of a still larger population, we may assume that the conclusion drawn from the experiment holds for this larger population also. If we wish to make this argument, however, it will have to be done on the basis of logical considerations apart from the test of significance. We have not randomly sampled this larger population.

The problems of random sampling from finite populations that are faced by the investigator in the behavioral sciences are, however, not peculiar to these sciences alone. They are faced by other research workers in other fields as well. An industrial engineer, testing the tensile strength of thread or some other material in a plant, cannot randomly sample an infinite population. The sample of thread he studies can be a random sample only of the population of thread already produced, and this is a finite population. Thread that has not yet been produced does not have an opportunity to appear in the sample. If the engineer argues that the sample at hand is representative of the thread that is still to be produced, he must do so on other than statistical grounds.

The same is true of the agronomist who conducts research with a variety of corn grown in a particular agricultural research station. He may have a random sample drawn from a limited population of plots of ground in, say, Iowa, but if it is argued that these plots are comparable to those in Washington or Michigan, this must be done on some other basis than that of random selection.

When a biologist selects from his laboratory two groups of rats to test two different drugs, he may randomly sample only the population of rats represented in his laboratory cages. He does not have a random sample from any larger population. If the results he obtains are assumed to hold for rats other than those represented in his own laboratory, this assumption must be made on some basis other than that of random selection from the larger population. The same considerations that apply to captive rats in a university laboratory also apply to captive sophomores in a university course in introductory psychology.

16.5 Experimental Conditions

An experimenter may confuse the notion of random selection of experimental units — subjects — with random selection of experimental conditions. In general, experimental conditions are fixed and cannot be considered as representing a random selection from a larger population of experimental conditions. For example, an experimenter may show subjects a motion picture film designed to influence their attitudes favorably toward members of a minority group. The film is prepared in two ways: Film 1 includes what is considered to be a highly emotional musical score, and Film 2 is prepared with what is considered to be a mildly emotional musical background. In all other respects, the two films are the same. It may be found that subjects viewing Film 1 show a greater change in attitude in the expected direction than subjects viewing Film 2. It should be obvious that the experimenter would be in error in concluding that "films with an emotional musical background are more influential in changing attitudes than those with a less emotional musical background." No sampling of attitudes, films, or musical scores has been investigated, and any conclusions drawn from the experiment logically apply only to the particular attitude, films, and musical scores actually investigated.

One must be careful in making generalizations that are not necessarily warranted on the basis of the actual conditions of an experiment. Suppose, for example, that after doing a limited experiment, a research worker concludes that "distributed practice is more effective than massed practice." This may be true for the particular materials investigated in the experiment. However, if the material which the subjects were required to learn consisted of nonsense syllables, it does not necessarily follow that the same results would be obtained with meaningful or other kinds of materials. Nor does the conclusion necessarily apply to other variations of distributed and massed practice than those investigated. If the subjects consisted only of children with IQ's of 125 and over, we do not know whether the same results would be found with older subjects or with those of lower intelligence.

In many psychological experiments, one of the experimental conditions involves the experimenter himself or one of his associates. For example, an investigator may be interested in students' scores on a scale designed to measure ethnocentrism when the test is administered by someone playing an "authoritarian" role and by someone playing a "democratic" role. It may be found that a significantly higher mean score is obtained by subjects tested by the "authoritarian." Obviously, any conclusions drawn about authoritarian and democratic leaders *in general* are not necessarily warranted by observing one authoritarian and one democratic leader *in particular*. No sampling of these types of leaders has been incorporated into the experimental design.

Another experimenter may have a sentence-completion test administered to one group of male subjects by a male and to another group of male subjects by a female. The results of the experiment may indicate that a significantly larger number of sexual responses are given when the test is administered by the female. This is an interesting result, but it does not necessarily mean that it would be obtained again with different female and male test administrators.

Similarly, when a condition of stress, anxiety, or ego involvement is created by the role-playing activities of the experimenter, it is as important to know about the experimenter and his personality as it is to know about the results he obtains. The experimenter himself is a part of the experimental conditions, and we need to know whether he, as well as the more physical aspects of the experimental conditions, can be duplicated. A physical variable, such as intensity of illumination, can be controlled and, within narrow limits, can be equated from one laboratory to another. It is relatively independent of any particular experimenter. But if an experimental condition is "anxiety" produced by the experimenter telling the subjects that they have failed on a task, the instructions given the subjects cannot so readily be separated from the individual who gives them. We should like to know whether the same instructions given by different individuals produce comparable results in the subjects under investigation.

16.6 Tests of Technique

Perhaps one of the most important aspects of experimental design is that known as a *test of technique*. In general, a test of technique involves a random subdivision of the experiment in such a way that the experiment itself is replicated or repeated. For example, we may be interested in performance of subjects under two different treatments or experimental conditions. Instead of randomly dividing n subjects into two groups of $n/2$ subjects each, they may be randomly divided into four groups of $n/4$ subjects each. The four groups are then randomly assigned with two groups for each treatment. We then have some basis for determining whether or not the results obtained under the treatments are repeatable. Even better would be a subdivision of the n subjects into 20 groups of $n/20$ subjects each. The 20 groups could then be assigned at random so that we have 10 pairs of groups. The various pairs would then constitute replications of the experiment and it would be possible to determine whether the direction of the 10 mean differences tends to be consistent.

If the experimental conditions involve role-playing activities on the part of the experimenter, it would also be possible to assign different experimenters to the various replications. If comparable results were obtained when different experimenters were used, we would have some basis for believing that it is the role played rather than the role-player that is of importance in determining the outcome of the experiment.

When an experiment is replicated, in the manner described, it is always possible to combine the various replications, if warranted, for the final analysis of the data.

16.7 Randomization Tests

Suppose that we have a set of n observations with an equal number obtained under each of k experimental conditions. It is theoretically possible to enumerate completely each of the possible different k sets, even though it would be laborious to do so if n were large. With a complete enumeration, it would also be possible to evaluate any property of each group, such as the arithmetic mean. If we have randomly assigned subjects to the experimental conditions, the particular set of means that we obtain, under the null hypothesis that the experimental conditions have no effect, would represent a random selection from the complete possible *population of sets*. The expected or average value of a given mean we know would then be the mean of the total number of observations. We also know, under the null hypothesis, that large differences in the k means would be more unusual or improbable than small differences.

For each possible set of k means, we could calculate the treatment sum of squares. This distribution would be exact and completely known. We consider only those treatment sums of squares that would be included in the 5 percent with the largest values, and we regard any one of these as significant. We would thus have an exact test of significance based only upon the notion of randomization. It is only the labor involved in developing the random distribution of statistics that prevents us from having exact tests of significance, free from certain assumptions, such as normality of distribution. By making certain assumptions, however, we can approximate the exact random distribution.

Suppose we have two experimental conditions with four observations obtained under each condition. The values of these observations are shown in Table 16.1, and it is apparent that the distribution is not normal but is rather that of a set of ranks. By methods described previously, the exact probability of $\bar{X}_1 - \bar{X}_2 \geq 3.0$ is known to be exactly equal to $4/70 = 0.057$, on the basis of random selection. For example, if we identify the eight observations by the letters A, B, C, D, E, F, G, and H, then we have $8!/4! = 1680$ permutations of the eight observations taken four at a time. If we ignore the order, then we have $1680/4! = 70$ combinations. Each of these combinations will yield a sum and a mean for Treatment 1. Because the sum of the eight observations is a constant, once we know the mean for Treatment 1, we also know the mean for Treatment 2. Only four of the 70 possible samples for Treatment 1 will result in a difference $\bar{X}_1 - \bar{X}_2 \geq 3.0$,

Table **16.1** — Values of a Variable X and X^2 Obtained under Different Experimental Conditions

	Condition 1		Condition 2	
	X	X^2	X	X^2
	6	36	5	25
	8	64	2	4
	7	49	1	1
	3	9	4	16
Σ	24	158	12	46

and those are the samples for which $\bar{X}_1 \geqslant 6.0$. These four samples for Treatment 1 are as follows:

$$8, 7, 6, 5, \qquad 8, 7, 6, 3,$$
$$8, 7, 6, 4, \qquad 8, 7, 5, 4,$$

Now let us apply a test of significance to the data given in Table 16.1, under various assumptions. We can then see how well the probability obtained from the test of significance agrees with the exact probability. Suppose, first of all, that we assume the eight observations constitute a population. We thus have a known population variance which, for a set of ranks, is

$$\sigma^2 = \frac{8^2 - 1}{12} = 5.25$$

and the population standard deviation is $\sigma = \sqrt{5.25} = 2.29$.

If we take a random sample of four observations from the population of eight, the standard error of the mean, using a finite population correction factor, will be

$$\sigma_{\bar{x}} = \sqrt{\frac{8-4}{8-1}} \times \frac{2.29}{\sqrt{4}} = 0.866$$

Then, since the population mean is known to be equal to 4.5, we have

$$Z = \frac{6.0 - 4.5}{0.866} = 1.73$$

If we interpret Z as a normal deviate, then we find, from the table of the normal curve, that the probability of Z equal to or greater than 1.73 is 0.0418. This is the probability of $\bar{X}_1 \geqslant 6.0$ and, therefore, of $\bar{X}_1 - \bar{X}_2 \geqslant 3.0$. Introducing a correction factor for discontinuity, we have

$$Z = \frac{5.875 - 4.5}{0.866} = 1.59$$

and the table of the normal curve gives 0.056 as the probability of Z equal to or greater than 1.59. The probability thus obtained by assuming normality of distribution is in very good agreement with the exact probability of 0.057.

Suppose that we now apply, as we most often would, a t test to the data of Table 16.1. In this instance, we shall not assume that the population mean and population variance are known. To obtain an estimate of the unknown population variance, we first calculate

$$\Sigma (X_1 - \bar{X}_1)^2 = 158 - \frac{(24)^2}{4} = 14$$

and

$$\Sigma (X_2 - \bar{X}_2)^2 = 46 - \frac{(12)^2}{4} = 10$$

The standard error of the difference between the two means will then be equal to

$$s_{\bar{X}_1 - \bar{X}_2} = \sqrt{\left(\frac{14 + 10}{4 + 4 - 2}\right)\left(\frac{1}{4} + \frac{1}{4}\right)} = 1.414$$

We then obtain

$$t = \frac{6.0 - 3.0}{1.414} = 2.12$$

with 6 d.f. By rough interpolation into the table of t, we find that the probability of t equal to or greater than 2.12 is 0.04. For the situation described, and under the null hypothesis tested, we can interpret this as the probability of $\bar{X}_1 - \bar{X}_2 \geqslant 3.0$, when the null hypothesis is true. The t test thus also provides a not too inaccurate estimate of the exact probability given by the randomization test.

In general it would seem that in applying tests of significance involving assumptions other than that of randomization, to experimental data, we shall be closest to the realities of the situation if we regard these tests as approximations of the exact probabilities provided by the corresponding complete randomization test for the observations actually made. The only population involved in the randomization test is that generated by the process of randomization for the particular observations made and which represent the outcome of the experiment. The only necessary condition is that the experimental conditions be assigned at random to the subjects, or vice versa. In using t, F, or χ^2 to approximate the probabilities of the randomization test, the assumptions we make are more in the nature of convenient computational devices rather than anything else.

In particular, if we have an adequate and equal number of observations for each of the treatment groups, the probabilities obtained from the tests of significance based on these distributions should be very good approximations of those that we would have obtained if we had made a randomization test.

REVIEW OF TERMS

error of measurement	randomization test
finite population	reliable measuring instrument
normal-distribution free test	systematic error
random error	test of technique

QUESTIONS AND PROBLEMS

16.1 Would you agree that, in general, the populations sampled by research workers are finite populations? Explain the basis of your answer.

16.2 Give an example such that it might be possible to regard the treatments or experimental conditions used in an experiment as a random sample from some larger population of experimental conditions. Explain your answer.

16.3 At the conclusion of an experiment, it is stated that "anxious subjects have higher levels of aspiration than nonanxious subjects." What are some of the qualifications that should be made with respect to this conclusion; that is, what questions would you want to put to the research worker responsible for the statement?

16.4 Discuss the problems involved when an experimenter himself is part of the experimental conditions.

16.5 Define a randomization test. What is the logical basis of such a test?

16.6 Explain what is meant by a test of technique. Of what value is a test of technique?

16.7 Why is it desirable, in an experiment, to have as reliable measurements as possible of a dependent variable?

16.8 Will a constant error have any influence on (a) the mean and (b) the variance of a set of measurements? Explain your answer.

Answers to the Problems

Chapter 1

1.1   (a) 0.0833   (c) 0.0714   (e) 0.0667   (g) 0.1000
     (b) 0.0400   (d) 0.0068   (f) 0.2500   (h) 0.0133

1.2   (a) 0.1250   (c) 12.5000
     (b) 1.2500   (d) 125.0000

1.3   (a) $k = 1$ or 2   (d) $k = 7, 8,$ or 9
     (b) $k = 1, 2, 3,$ or 4   (e) $k = 4$ or 5
     (c) $k = 5, 6, 7, 8,$ or 9   (f) $k = 1, 2, 3,$ or 4

1.4   (a) 221   (e) 1.85
     (b) 371   (f) 2.26
     (c) 395   (g) 3.76
     (d) 333   (h) 4.99

1.5   (a) 14.8324   (i)   0.221
     (b) 1.48324   (j) 37.1
     (c) 0.148324   (k)   3.71
     (d) 64   (l) 23.1
     (e)   6.4   (m)   2.31
     (f)   0.64   (n) 39.6
     (g) 22.1   (o)   3.96
     (h)   2.21   (p)   0.396.

Chapter 2

2.1   (a) continuous   (f) discrete
     (b) continuous   (g) continuous
     (c) continuous   (h) discrete
     (d) discrete   (i) discrete
     (e) discrete   (j) discrete

2.2   All

2.3   (a) 28.35   (f) 133.46
      (b) 2.68   (g) 18.46
      (c) 32.36   (h) 23.79
      (d) 1.85   (i) 22.68
      (e) 16.36   (j) 13.80

Chapter 3

3.1   (a) 54   (d) 62
      (b) 43.5   (e) 60–64
      (c) 29.5

3.7   Limits   Midpoint
      (a)   19.5–23.5   21.5
      (b)   8.5–12.5   10.5
      (c) 119.5–124.5   122.0
      (d) 0.095–0.205   0.150
      (e)   2.45–3.55   3.0
      (f) 0.845–0.955   0.90

3.8   (a) 2   (d) 10
      (b) 6   (e) 3
      (c) 4

3.9   Limits   Midpoint
      (a)   32–33   32.5
      (b)   12–17   14.5
      (c)   12–15   13.5
      (d)   10–19   14.5
      (e)   45–47   46.0

Chapter 4

4.1   (a) $\bar{X} = 60.10$
      (b) Mdn = 60.46

4.2 $\bar{X} = \Sigma fX/n = 160/200 = 0.8$

4.3 $\Sigma X = 10(10+1)/2 = 55$
      $\bar{X} = 55/10 = (10+1)/2 = 5.5$

4.4   (a) $\bar{X} = 4.0$ (c) $\bar{X} = 6.0$
      (b) $\bar{X} = 5.0$ (d) $\bar{X} = 7.0$

4.5   (a) $\bar{X} = 24.25$
      (b) Mdn = 24.3

4.7   (a) $\bar{X} = 8.0$
      (b) $\bar{X} = 12.0$
      (c) $\bar{X} = 16.0$

4.8 $\bar{X} = 8.44$ rounded

4.9 $\bar{X} = 0.67$ rounded

4.10 (a) $\bar{X} = 3.0$
    (b) $\bar{X} = 2.0$
    (c) $\bar{X} = 0$

4.11 We have $X + a$. Then summing and dividing by n, we obtain

$$\frac{\Sigma X + na}{n} = \bar{X} + a$$

4.12 (a) $\bar{X} = 2.0$
    (b) $\bar{X} = 1.0$

4.13 We have bX. Then summing and dividing by n, we obtain

$$\frac{b\Sigma X}{n} = b\bar{X}$$

4.15 We have

$$\frac{\Sigma(X - \bar{X})}{n} = \frac{\Sigma X - n\bar{X}}{n} = 0$$

because $\bar{X} = \Sigma X / n$ and, consequently, $n\bar{X} = \Sigma X$.

4.17 $\bar{X} = 7.0$

4.18 $C_{50} = 5.0$

4.19 $\bar{X} = 5.0$

4.20 (a) 20.0   (d) 12.0
    (b) 8.0   (e) 0
    (c) 5.0

Chapter 5

5.1 (a) $s^2 = 103.42$
    (b) $s = 10.2$

5.2 (a) $\Sigma(X - \mu)^2 = (N^3 - N)/12 = 720/12 = 60$
    (b) $\sigma^2 = \Sigma(X - \mu)^2/N = 60/9 = (81 - 1)/12 = 20/3 = 6.67$

5.3 (a) $\mu = \Sigma fX/N = 160/200 = 0.8$
    (b) $\sigma^2 = \Sigma F(X - \mu)^2/N = (0.8)(0.2) = 0.16$

5.6 (a) $s^2 = 4.0$ (c) $s^2 = 4.0$
    (b) $s^2 = 4.0$ (d) $s^2 = 4.0$

5.7 $s^2 = 4.0$

5.8 (a) $s^2 = 16.0$
    (b) $s^2 = 36.0$
    (c) $s^2 = 64.0$

5.9 (a) $s^2 = (1/2)^2(4) = 1.0$
      (b) $s^2 = (1/4)^2(4) = 0.25$

5.10 We have bX. We know that the mean of bX will be $b\bar{X}$. Then $bX - b\bar{X} = b(X - \bar{X})$ and

$$\frac{b^2\Sigma(X-\bar{X})^2}{n-1} = b^2 s^2$$

5.11 We know that

$$\frac{X}{b} = \frac{1}{b}X$$

and from Problem 5.10 we have

$$\frac{\left(\frac{1}{b}\right)^2 \Sigma(X-\bar{X})^2}{n-1} = \left(\frac{1}{b}\right)^2 s^2$$

5.12 $\mu = 13.0$ and $\sigma = 7.211$

5.15 $\bar{X} = 11.0$ and $s = 2.0$

5.17 $\mu = 37.0$ and $\sigma = 21.07$

5.18 $\bar{X} = 5.0$ and $s = 3.46$

5.19 The proof can be accomplished in a number of ways of which we give but one.

$$\frac{\Sigma(X-\mu)^2}{N} = \frac{\Sigma X^2 - 2\mu\Sigma X + N\mu^2}{N}$$

But $\Sigma X = \Sigma X^2 = F_1$ and $\mu = P = F_1/N$, so that

$$\frac{\Sigma(X-\mu)^2}{N} = P - 2P^2 + P^2$$
$$= P(1 - 2P + P)$$
$$= P(1 - P)$$
$$= PQ$$

5.20 $\Sigma(X-\bar{X})^2 = \Sigma X^2 - 2\bar{X}\Sigma X + n\bar{X}^2$
$\qquad = \Sigma X^2 - 2n\bar{X}^2 + n\bar{X}^2$
$\qquad = \Sigma X^2 - n\bar{X}^2$
$\qquad = \Sigma X^2 - (\Sigma X)^2/n$

5.21 $\mu = \Sigma X/N$ and for a binomial population $\Sigma X = F_1$ so that

$$\Sigma X/N = F_1/N = P$$

5.25 (a) $s^2 = 16.0$ and $s = 4.0$
      (b) $s^2 = 4.0$ and $s = 2.0$
      (c) $s^2 = 1.0$ and $s = 1.0$
      (d) $s^2 = 4.0$ and $s = 2.0$
      (e) $s^2 = 1.0$ and $s = 1.0$

5.26 See the proof for 5.20

Chapter 6

6.1 (a) $r = -0.73$
    (b) $\bar{X} = 8.0$
    (c) $\bar{Y} = 9.0$
    (d) $b_Y = -0.73$
    (e) $b_X = -0.73$
    (f) $s_{Y \cdot X} = 3.75$

6.2 (a) $\tilde{Y} = 13.38$
    (b) $\tilde{Y} = 9.00$
    (c) $\tilde{Y} = 4.62$

6.3 (a) $\tilde{X} = 10.92$
    (b) $\tilde{X} = 8.00$
    (c) $\tilde{X} = 4.35$

6.11 $\tilde{z}_Y = 1.2$

6.12 $r = 0.30$

6.13 $b_X = 0.75$

6.14 $s_Y = 16.0$

6.15 (a) $\bar{z} = 0$
    (b) $s_z^2 = 1.0$

6.18 $b_Y = b_X = 0.80$

6.21 (a) $a = 2$ $b = 1$
    (b) $a = 0$ $b = 3$
    (c) $a = -3$ $b = 2$
    (d) $a = 1$ $b = -2$
    (e) $a = -5$ $b = 2$

6.22 (a) $c_{XY} = -0.50$ (b) $r = -0.50$

6.23 $r = 0.80$

6.24 (a) $r = 1.00$
    (b) $b_Y = b$
    (c) $s_{Y \cdot X} = 0$

6.25 $c = 10.0$

6.29 (a) $\bar{Z} = \dfrac{\Sigma Z}{n} = \dfrac{1}{s} \dfrac{\Sigma (X - \bar{X})}{n} = \dfrac{1}{s} \dfrac{\Sigma X - n\bar{X}}{n} = 0$

because we have previously proved that $\Sigma X = n\bar{X}$.
    (b) If $\bar{Z} = 0$, then

$$s_Z^2 = \frac{\Sigma (Z - \bar{Z})^2}{n - 1} = \frac{\Sigma Z^2}{n - 1}$$

and

$$s_Z{}^2 = \frac{\dfrac{\Sigma(X-\bar{X})^2}{s^2}}{n-1} = \frac{\Sigma(X-\bar{X})^2}{s^2(n-1)} = \frac{s^2}{s^2} = 1.0$$

Therefore $s_Z = \sqrt{s_Z{}^2} = 1.0$

6.31 $r = 0.60$

Chapter 7

7.1 (a) 720   (d) 15
     (b) 120   (e) 60
     (c) 20

7.2 (a) 4/7   (d) 6/7
     (b) 2/7   (e) 5/7
     (c) 1/7   (f) 3/7

7.3 (a) 8/49   (d) 2/49
     (b) 4/49   (e) 4/49
     (c) 8/49   (f) 2/49

7.4 (a) 16/49
     (b) 4/49
     (c) 1/49
     (d) We note that the above three samples and the six samples of Problem 7.3 exhaust the possibilities. Only one sample will result in two A's and the probability of this sample is 16/49. In addition, the following samples will result in one A: AB, AC, BA, and CA. Because these outcomes are mutually exclusive, the desired probability is

$$\frac{16}{49} + \frac{8}{49} + \frac{4}{49} + \frac{8}{49} + \frac{4}{49} = \frac{40}{49}$$

(e) In the same manner in which we obtained the answer to (d), we find that the probability is 24/49.
(f) In the same manner in which we obtained the answer to (d), we find that the probability is 13/49.

7.5 The number of ways in which three correct and one incorrect answer can be obtained in a set of 4 questions is

$$\frac{4!}{3!1!}$$

The probability for each of these ways is $(1/4)^3(3/4)$. The desired probability is, therefore,

$$\frac{4!}{3!1!}(1/4)^3(3/4) = 3/64$$

7.6 (a) 1   (d) 10
     (b) 5   (e) 5
     (c) 10   (f) 1

7.7 (a) 1/243   (d) 80/243
     (b) 10/243   (e) 80/243
     (c) 40/243   (f) 32/243

7.8 (a) 1/64
     (b) 6/64
     (c) 22/64

7.9 (a) 10/25   (e) 20/25
     (b) 15/25   (f) 5/15
     (c) 5/25   (g) 5/10
     (d) 5/25

7.10

$$\frac{10!}{3!7!} = 120$$

7.11

$$\frac{6!}{3!3!} \times \frac{4!}{2!2!} = 120$$

7.12 We have $6! = 720$ different orders. With two subjects for each order, we need $(2)(720) = 1440$ subjects.

7.13 105

7.14 (a) 10/25   (g) 12/25
      (b) 15/25   (h) 13/25
      (c) 7/12   (i) 7/10
      (d) 5/12   (j) 3/10
      (e) 3/13   (k) 5/15
      (f) 10/13   (l) 10/15

   (m) $P(A \text{ and } D) = \frac{10}{25} \times \frac{3}{10} = \frac{13}{25} \times \frac{3}{13} = \frac{3}{25}$

   (n) $P(A \text{ or } D) = \frac{10}{25} + \frac{13}{25} - \frac{3}{25} = \frac{20}{25}$

      (o) 10/25   (r) 15/25
      (p) 18/25   (s) 5/25
      (q) 7/25   (t) 22/25

7.15 You should note the relationship between the solution of this problem and formula (7.6) in the text.

$$\frac{15!}{5!10!} \times \frac{10!}{5!5!} \times \frac{5!}{5!0!} = \frac{15!}{5!5!5!}$$

7.16 $5! = 120$

7.17 We can select two from the set of five in 10 ways. Only one of the ways is correct. The desired probability is, therefore, 1/10. We can also regard the problem in this way: If we name one object, the probability of a correct choice is 2/5. Given that a correct choice has been made, the conditional probability that the second choice will be correct is 1/4. The probability that both choices will be correct is, therefore, $(2/5)(1/4) = 1/10$.

7.21 45

7.22 (a) $P = 12/20 = 0.6$
      (b) $P = 8/20 = 0.4$
      (c) $P = 6/20 = 0.3$
      (d) $P = 2/20 = 0.1$
      (e) $P = 0.9$

7.23 (a) $P = 6/60 = 0.1$
      (b) $P = 36/60 = 0.6$
      (c) $P = 18/60 = 0.3$
      (d) $P = 0$

7.24 24

7.25 336

7.26 1680

7.27 (a) 90   (c) 15
      (b) 480   (d) 1

Chapter 8

8.1 (a) $\mu = 3.0$
     (b) $\sigma^2 = 1.2$

8.3 (b) $\mu_{\bar{X}} = 3.0$
     (c) $\sigma_{\bar{X}}^2 = 0.60$
     (d) $\sigma_{\bar{X}} = 0.775$

8.5 $n = 270$

8.6 $\sigma_{\bar{X}} = 0.730$

8.7 (a) $\sigma_{\bar{X}} = 0.754$
     (b) $\sigma_{\bar{X}} = 0.761$
     (c) $\sigma_{\bar{X}} = 0.767$

8.8 (a) $\mu_D = 0$
     (b) $\sigma_D = 7.07$

8.9 (a) $\mu_T = 45$
     (b) $\sigma_T = 8.66$

8.11 (a) $\mu_D = 0$
(b) $\sigma_D^2 = 2.4$

8.12 (a) $\mu_T = 6.0$
(b) $\sigma_T^2 = 2.4$

8.13 (a) $P = 1/2$
(b) $P = 0$
(c) $P = 1/2$

8.14 (a) $\mu_T = 1.0$
(b) $\sigma_T^2 = 1.0$

8.15 (a) $\mu_T = 1.0$
(b) $\sigma_T^2 = 1.0$

8.16 (a) $\mu_T = 1.0$
(b) $\sigma_T^2 = 1.0$

8.17 (a) $\mu_T = 1.0$
(b) $\sigma_T^2 = 1.0$

8.18 $\sigma_{\bar{X}} = 1.5$

8.19 $\mu = 0.8$ and $\sigma = 0.6$

8.20 16

8.21 $\mu_T = 35.0$ and $\sigma_T^2 = 29.17$

8.22 If $n = 2$, then

$$\bar{X} = \frac{X_1 + X_2}{2}$$

and

$$\Sigma(X - \bar{X})^2 = \left(X_1 - \frac{X_1 + X_2}{2}\right)^2 + \left(X_2 - \frac{X_1 + X_2}{2}\right)^2$$

$$= \frac{(2X_1 - X_1 - X_2)^2 + (2X_2 - X_1 - X_2)^2}{4}$$

$$= \frac{(X_1 - X_2)^2 + (X_2 - X_1)^2}{4}$$

$$= \frac{2(X_1 - X_2)^2}{4}$$

$$= \frac{(X_1 - X_2)^2}{2}$$

and, because $n - 1 = 1$, we have

$$s^2 = \frac{(X_1 - X_2)^2}{2}$$

Chapter 9

9.1 (a) 45
     (b) $(1/3)^8(2/3)^2$
     (c) $45(1/3)^8(2/3)^2$

9.3 (a) $\mu_T = 1.5$
     (b) $\sigma_T^2 = 1.125$

9.4 (a) $\mu_T = 6.0$
     (b) $\sigma_T^2 = 4.0$

9.5 (a) $\mu_T = 5.0$
     (b) $\sigma_T^2 = 2.5$

9.7 (b) $\mu_T = 1.5$
     (c) $\sigma_T^2 = 0.75$

9.8 (a) 1
     (b) 10
     (c) 45
     (d) 56/1024

9.9 (a) $\mu_T = 6.0$
     (b) $\sigma_T^2 = 5.0$
     (c) $\dfrac{36!}{22!14!}(1/6)^{22}(5/6)^{14}$
     (d) $\dfrac{36!}{5!31!}(1/6)^5(5/6)^{31}$

9.10 (a) $(1/2)^4$
     (b) $(1/4)^4$
     (c) $\dfrac{4!}{2!2!}(1/4)^2(3/4)^2$
     (d) $\dfrac{4!}{2!2!}(1/13)^2(12/13)^2$
     (e) $\dfrac{4!}{1!3!}(1/13)(12/13)^3$

9.12 $\mu_T = 30$ and $\sigma_T = 4.90$

9.13 $(1/5)^6 + 6(1/5)^5(4/5) + 15(1/5)^4(4/5)^2 + 20(1/5)^3(4/5)^3 + 15(1/5)^2(4/5)^4$
     $+ 6(1/5)(4/5)^5 + (4/5)^6$

9.14 (a) $\mu = 1/3$ and $\sigma^2 = 2/9$
     (b) $\mu_T = 2/3$ and $\sigma_T^2 = 2/9$

9.15 $\mu_T = 9/4$ and $\sigma_T^2 = 9/16$

9.16 $\mu_T = 1.2$ and $\sigma_T^2 = 0.36$

Chapter 10

10.2 (a) 60.5   (e) 24.5 and 75.5
      (b) 59.5   (f) 25.5 and 75.5
      (c) 24.5   (g) 24.5 and 74.5
      (d) 30.5   (h) 25.5 and 74.5

10.3 (a) $X \geqslant 67$ (b) $X \geqslant 69$

10.4 $Z = 2.24$ and $P(Z \geqslant 2.24) = 0.0125$
      With a correction for discontinuity, $Z = 2.01$ and $P(Z \geqslant 2.01) = 0.0222$

10.5 (a) $n = 15$ trials
      (b) $Z = 2.32$ and $P(Z \geqslant 2.32) = 0.0102$
      (d) $Z = 2.13$ and $P(Z \geqslant 2.13) = 0.0166$

10.6 (a) $P = 0.05$
      (b) $\sigma_{\bar{X}} = 0.775$
      (c) $Z = 1.94$ and $P(Z \geqslant 1.94) = 0.0262$
      (e) $Z = 1.61$ and $P(Z \geqslant 1.61) = 0.0537$

10.7 $Z = 2.50$ and $P(Z \geqslant 2.50) = 0.0062$

10.8 $Z = 2.68$ and $P(Z \geqslant 2.68) = 0.0037$

10.9 (a) $\mu_T = 6.0$
      (b) $\sigma_T^2 = 5.0$
      (c) $Z = 1.12$ and $P(Z \geqslant 1.12) = 0.1314$

10.10 (a) $\mu_T = 630.0$ and $\sigma_T = 29.91$
      (b) $Z = -0.99$ and $P(Z \leqslant -0.99) = 0.1611$

10.11 (a) $\mu_T = 60.0$ and $\sigma_T = 6.0$
      (b) $Z = -2.58$ and $P(Z \leqslant -2.58) = 0.0049$

10.12 (a) $\mu = 13.0$ and $\sigma^2 = 52.0$
      (b) $\mu_T = 65.0$ and $\sigma_T^2 = 216.67$
      (c) $Z = 0.65$ and $P(Z \geqslant 0.65) = 0.2578$

10.13 (a) $\mu_T = 12$ and $\sigma_T = 3.0$
      (b) $Z = 1.5$ and $P(Z \geqslant 1.5) = 0.0668$

10.14 (a) $\mu_T = 6.0$ and $\sigma_T = 2.0$
      (b) $Z = 2.75$ and $P(Z \geqslant 2.75) = 0.0030$

Chapter 11

11.1 (a) $s_{\bar{X}} = 2.0$
      (b) $c_1 = 13.872$ and $c_2 = 22.128$

11.2 (c) $\mu = 3.0$

11.3 $t = 3.16$

11.4 (c) $\mu = 4.5$

11.5 $t = 2.33$

11.6 $c_1 = 26.98$ and $c_2 = 33.02$

11.12 $t = 2.53$ with 4 d.f.

11.13 (a) $s_{\bar{x}} = 1.6$
     (b) $c_1 = 69.70$ and $c_2 = 76.30$

11.14 $t = 3.00$ with 8 d.f.

11.15 (a) $s_{\bar{x}_1 - \bar{x}_2} = 2.0$
     (b) $t = 1.5$ with 18 d.f.
     (c) 40
     (d) $t = 3.0$ with 38 d.f.

Chapter 12

12.1 $F = 11.79$

12.2   (a) $t = 2.535$
     (b) $F = 6.43$

12.6   (a) $t = 5.18$
     (b) $t = 2.92$
     (c) $t = 0$

12.7 $F = 3.25$

12.10 (a) $t = 3.0$ with 8 d.f.
     (b) $F = 9.0$ with 1 and 8 d.f.

12.11 $F = 3.5$ with 3 and 8 d.f.

12.12 $F = 5.0$ with 4 and 20 d.f.

12.13 $F = 1.00$ with 1 and 8 d.f.

12.14 $t = 1.00$ with 1 and 8 d.f.

Chapter 13

13.1   (a) $\chi^2 = 12.00$
     (b) 60.5 and 39.5

13.2 $\chi^2 = 8.75$

13.3 $\chi^2 = 14.0$

13.4 $\chi^2 = 7.14$

13.7   (a) $\chi^2 = 4.00$ with 1 d.f.
     (b) $Z = 2.00$

13.8 $\chi^2 = 20.0$ with 2 d.f.

13.9   (a) $\chi^2 = 8.41$
     (b) $Z = 2.9$

Chapter 14

14.1 (a) $r = 0.33$
(b) $\chi^2 = 10.89$

14.2 $C = 0.35$

14.3 (a) $r = 0.24$
(b) $F = 1.49$

14.4 (a) $\eta_{YX} = 0.71$
(b) $F = 6.17$

14.5 (a) $r = 0.79$
(b) $t = 3.65$

14.6 $t = 1.69$

14.7 $r = 0.56$

14.8 $t = 2.25$ with 9 d.f.

14.9 $C = 0.37$

14.10 $r = 0.09$

14.11 $r = 0.80$

14.12 $r = 0.80$

14.13 $r = 0.20$

14.14 $r = 0.20$

14.15 $r^2 = 0.75$

14.16 $r^2 = 0.75$

Chapter 15

15.1 $r_{kk} = 0.89$

15.2 $r_{XY} = 0.72$

15.3 $r_{Y_1Y_2} = 1.00$

15.4 $r_{kk} = 0.75$

15.5 Solve

$$k = n\left[\frac{r_{kk}(1 - r_{nn})}{r_{nn}(1 - r_{kk})}\right]$$

with $n = 10$, $r_{kk} = 0.90$, and $r_{nn} = 0.60$. Then $k = 60$. Alternatively,

$$0.9 = \frac{m(0.6)}{1 + (m - 1)0.6}$$

and solving for m, we have

$$m = 0.36/0.06 = 6$$

Then

$$6 = \frac{k}{10} \text{ and } k = 60$$

15.7 (a) $\bar{X} = 13.61$
(b) K-R 20 = 0.87

15.8 K-R 20 = 0.92

15.9 (a) The mean will be increased by two points.
(b) The standard deviation will remain the same.
(c) The covariance will remain the same.
(d) The correlation coefficient will remain the same.
(e) The standard error of the mean will remain the same.

APPENDIX

TABLE I. *Table of Random Numbers**

COLUMN NUMBER

1st Thousand

Row	00000 01234	00000 56789	11111 01234	11111 56789	22222 01234	22222 56789	33333 01234	33333 56789
00	23157	54859	01837	25993	76249	70886	95230	36744
01	05545	55043	10537	43508	90611	83744	10962	21343
02	14871	60350	32404	36223	50051	00322	11543	80834
03	38976	74951	94051	75853	78805	90194	32428	71695
04	97312	61718	99755	30870	94251	25841	54882	10513
05	11742	69381	44339	30872	32797	33118	22647	06850
06	43361	28859	11016	45623	93009	00499	43640	74036
07	93806	20478	38268	04491	55751	18932	58475	52571
08	49540	13181	08429	84187	69538	29661	77738	09527
09	36768	72633	37948	21569	41959	68670	45274	83880
10	07092	52392	24627	12067	06558	45344	67338	45320
11	43310	01081	44863	80307	52555	16148	89742	94647
12	61570	06360	06173	63775	63148	95123	35017	46993
13	31352	83799	10779	18941	31579	76448	62584	86919
14	57048	86526	27795	93692	90529	56546	35065	32254
15	09243	44200	68721	07137	30729	73756	09298	27650
16	97957	35018	40894	88329	52230	82521	22532	61587
17	93732	59570	43781	98885	56671	66826	95996	44569
18	72621	11225	00922	68264	35666	59434	71687	58167
19	61020	74418	45371	20794	95917	37866	99536	19378
20	97839	85474	33055	91718	45473	54144	22034	23000
21	89160	97192	22232	90637	35055	45489	88438	16361
22	25966	88220	62871	79265	02823	52862	84919	54883
23	81443	31719	05049	54806	74690	07567	65017	16543
24	11322	54931	42362	34386	08624	97687	46245	23245

* Table I is reproduced from M. G. Kendall and B. B. Smith. Randomness and random sampling numbers. *Journal of the Royal Statistical Society*, **101** (1938), 147–166, by permission of the Royal Statistical Society.

TABLE I. *Table of Random Numbers*—*Continued*

COLUMN NUMBER

4th Thousand

Row	00000 01234	00000 56789	11111 01234	11111 56789	22222 01234	22222 56789	33333 01234	33333 56789
00	02490	54122	27944	39364	94239	72074	11679	54082
01	11967	36469	60627	83701	09253	30208	01385	37482
02	48256	83465	49699	24079	05403	35154	39613	03136
03	27246	73080	21481	23536	04881	89977	49484	93071
04	32532	77265	72430	70722	86529	18457	92657	10011
05	66757	98955	92375	93431	43204	55825	45443	69265
06	11266	34545	76505	97746	34668	26999	26742	97516
07	17872	39142	45561	80146	93137	48924	64257	59284
08	62561	30365	03408	14754	51798	08133	61010	97730
09	62796	30779	35497	70501	30105	08133	00997	91970
10	75510	21771	04339	33660	42757	62223	87565	48468
11	87439	01691	63517	26590	44437	07217	98706	39032
12	97742	02621	10748	78803	38337	65226	92149	59051
13	98811	06001	21571	02875	21828	83912	85188	61624
14	51264	01852	64607	92553	29004	26695	78583	62998
15	40239	93376	10419	68610	49120	02941	80035	99317
16	26936	59186	51667	27645	46329	44681	94190	66647
17	88502	11716	98299	40974	42394	62200	69094	81646
18	63499	38093	25593	61995	79867	80569	01023	38374
19	36379	81206	03317	78710	73828	31083	60509	44091
20	93801	22322	47479	57017	59334	30647	43061	26660
21	29856	87120	56311	50053	25365	81265	22414	02431
22	97720	87931	88265	13050	71017	15177	06957	92919
23	85237	09105	74601	46377	59938	15647	34177	92753
24	75746	75268	31727	95773	72364	87324	36879	06802

* Table I is reproduced from M. G. Kendall and B. B. Smith. Randomness and random sampling numbers. *Journal of the Royal Statistical Society*, **101** (1938), 147–166, by permission of the Royal Statistical Society.

TABLE I. *Table of Random Numbers*—*Concluded*

COLUMN NUMBER

Row	00000 01234	00000 56789	11111 01234	11111 56789	22222 01234	22222 56789	33333 01234	33333 56789
				5th Thousand				
00	29935	06971	63175	52579	10478	89379	61428	21363
01	15114	07126	51890	77787	75510	13103	42942	48111
02	03870	43225	10589	87629	22039	94124	38127	65022
03	79390	39188	40756	45269	65959	20640	14284	22960
04	30035	06915	79196	54428	64819	52314	48721	81594
05	29039	99861	28759	79802	68531	39198	38137	24373
06	78196	08108	24107	49777	09599	43569	84820	94956
07	15847	85493	91442	91351	80130	73752	21539	10986
08	36614	62248	49194	97209	92587	92053	41021	80064
09	40549	54884	91465	43862	35541	44466	88894	74180
10	40878	08997	14286	09982	90308	78007	51587	16658
11	10229	49282	41173	31468	59455	18756	08908	06660
12	15918	76787	30624	25928	44124	25088	31137	71614
13	13403	18796	49909	94404	64979	41462	18155	98335
14	66523	94596	74908	90271	10009	98648	17640	68909
15	91665	36469	68343	17870	25975	04662	21272	50620
16	67415	87515	08207	73729	73201	57593	96917	69699
17	76527	96996	23724	33448	63392	32394	60887	90617
18	19815	47789	74348	17147	10954	34355	81194	54407
19	25592	53587	76384	72575	84347	68918	05739	57222
20	55902	45539	63646	31609	95999	82887	40666	66692
21	02470	58376	79794	22482	42423	96162	47491	17264
22	18630	53263	13319	97619	35859	12350	14632	87659
23	89673	38230	16063	92007	59503	38402	76450	33333
24	62986	67364	06595	17427	84623	14565	82860	57300

* Table I is reproduced from M. G. Kendall and B. B. Smith. Randomness and random sampling numbers. *Journal of the Royal Statistical Society*, **101** (1938), 147–166, by permission of the Royal Statistical Society.

TABLE II. *Table of Squares, Square Roots, and Reciprocals of Numbers from 1 to 1000**

N	N^2	$\sqrt{N}$	$1/N$	N	N^2	$\sqrt{N}$	$1/N$
1	1	1.0000	1.000000	41	1681	6.4031	.024390
2	4	1.4142	.500000	42	1764	6.4807	.023810
3	9	1.7321	.333333	43	1849	6.5574	.023256
4	16	2.0000	.250000	44	1936	6.6332	.022727
5	25	2.2361	.200000	45	2025	6.7082	.022222
6	36	2.4495	.166667	46	2116	6.7823	.021739
7	49	2.6458	.142857	47	2209	6.8557	.021277
8	64	2.8284	.125000	48	2304	6.9282	.020833
9	81	3.0000	.111111	49	2401	7.0000	.020408
10	100	3.1623	.100000	50	2500	7.0711	.020000
11	121	3.3166	.090909	51	2601	7.1414	.019608
12	144	3.4641	.083333	52	2704	7.2111	.019231
13	169	3.6056	.076923	53	2809	7.2801	.018868
14	196	3.7417	.071429	54	2916	7.3485	.018519
15	225	3.8730	.066667	55	3025	7.4162	.018182
16	256	4.0000	.062500	56	3136	7.4833	.017857
17	289	4.1231	.058824	57	3249	7.5498	.017544
18	324	4.2426	.055556	58	3364	7.6158	.017241
19	361	4.3589	.052632	59	3481	7.6811	.016949
20	400	4.4721	.050000	60	3600	7.7460	.016667
21	441	4.5826	.047619	61	3721	7.8102	.016393
22	484	4.6904	.045455	62	3844	7.8740	.016129
23	529	4.7958	.043478	63	3969	7.9373	.015873
24	576	4.8990	.041667	64	4096	8.0000	.015625
25	625	5.0000	.040000	65	4225	8.0623	.015385
26	676	5.0990	.038462	66	4356	8.1240	.015152
27	729	5.1962	.037037	67	4489	8.1854	.014925
28	784	5.2915	.035714	68	4624	8.2462	.014706
29	841	5.3852	.034483	69	4761	8.3066	.014493
30	900	5.4772	.033333	70	4900	8.3666	.014286
31	961	5.5678	.032258	71	5041	8.4261	.014085
32	1024	5.6569	.031250	72	5184	8.4853	.013889
33	1089	5.7446	.030303	73	5329	8.5440	.013699
34	1156	5.8310	.029412	74	5476	8.6023	.013514
35	1225	5.9161	.028571	75	5625	8.6603	.013333
36	1296	6.0000	.027778	76	5776	8.7178	.013158
37	1369	6.0828	.027027	77	5929	8.7750	.012987
38	1444	6.1644	.026316	78	6084	8.8318	.012821
39	1521	6.2450	.025641	79	6241	8.8882	.012658
40	1600	6.3246	.025000	80	6400	8.9443	.012500

* Portions of Table II have been reproduced from J. W. Dunlap and A. K. Kurtz. *Handbook of Statistical Nomographs, Tables, and Formulas*, World Book Company, New York (1932), by permission of the authors and publishers.

TABLE II. *Table of Squares, Square Roots, and Reciprocals of Numbers from 1 to 1000*—Continued*

N	N^2	$\sqrt{N}$	$1/N$	N	N^2	$\sqrt{N}$	$1/N$
81	6561	9.0000	.012346	121	14641	11.0000	.00826446
82	6724	9.0554	.012195	122	14884	11.0454	.00819672
83	6889	9.1104	.012048	123	15129	11.0905	.00813008
84	7056	9.1652	.011905	124	15376	11.1355	.00806452
85	7225	9.2195	.011765	125	15625	11.1803	.00800000
86	7396	9.2736	.011628	126	15876	11.2250	.00793651
87	7569	9.3274	.011494	127	16129	11.2694	.00787402
88	7744	9.3808	.011364	128	16384	11.3137	.00781250
89	7921	9.4340	.011236	129	16641	11.3578	.00775194
90	8100	9.4868	.011111	130	16900	11.4018	.00769231
91	8281	9.5394	.010989	131	17161	11.4455	.00763359
92	8464	9.5917	.010870	132	17424	11.4891	.00757576
93	8649	9.6437	.010753	133	17689	11.5326	.00751880
94	8836	9.6954	.010638	134	17956	11.5758	.00746269
95	9025	9.7468	.010526	135	18225	11.6190	.00740741
96	9216	9.7980	.010417	136	18496	11.6619	.00735294
97	9409	9.8489	.010309	137	18769	11.7047	.00729927
98	9604	9.8995	.010204	138	19044	11.7473	.00724638
99	9801	9.9499	.010101	139	19321	11.7898	.00719424
100	10000	10.0000	.010000	140	19600	11.8322	.00714286
101	10201	10.0499	.00990099	141	19881	11.8743	.00709220
102	10404	10.0995	.00980392	142	20164	11.9164	.00704225
103	10609	10.1489	.00970874	143	20449	11.9583	.00699301
104	10816	10.1980	.00961538	144	20736	12.0000	.00694444
105	11025	10.2470	.00952381	145	21025	12.0416	.00689655
106	11236	10.2956	.00943396	146	21316	12.0830	.00684932
107	11449	10.3441	.00934579	147	21609	12.1244	.00680272
108	11664	10.3923	.00925926	148	21904	12.1655	.00675676
109	11881	10.4403	.00917431	149	22201	12.2066	.00671141
110	12100	10.4881	.00909091	150	22500	12.2474	.00666667
111	12321	10.5357	.00900901	151	22801	12.2882	.00662252
112	12544	10.5830	.00892857	152	23104	12.3288	.00657895
113	12769	10.6301	.00884956	153	23409	12.3693	.00653595
114	12996	10.6771	.00877193	154	23716	12.4097	.00649351
115	13225	10.7238	.00869565	155	24025	12.4499	.00645161
116	13456	10.7703	.00862069	156	24336	12.4900	.00641026
117	13689	10.8167	.00854701	157	24649	12.5300	.00636943
118	13924	10.8628	.00847458	158	24964	12.5698	.00632911
119	14161	10.9087	.00840336	159	25281	12.6095	.00628931
120	14400	10.9545	.00833333	160	25600	12.6491	.00625000

* Portions of Table II have been reproduced from J. W. Dunlap and A. K. Kurtz. *Handbook of Statistical Nomographs, Tables, and Formulas*, World Book Company, New York (1932), by permission of the authors and publishers.

TABLE II. *Table of Squares, Square Roots, and Reciprocals
of Numbers from 1 to 1000*—Continued*

N	N^2	$\sqrt{N}$	$1/N$	N	N^2	$\sqrt{N}$	$1/N$
161	25921	12.6886	.00621118	201	40401	14.1774	.00497512
162	26244	12.7279	.00617284	202	40804	14.2127	.00495050
163	26569	12.7671	.00613497	203	41209	14.2478	.00492611
164	26896	12.8062	.00609756	204	41616	14.2829	.00490196
165	27225	12.8452	.00606061	205	42025	14.3178	.00487805
166	27556	12.8841	.00602410	206	42436	14.3527	.00485437
167	27889	12.9228	.00598802	207	42849	14.3875	.00483092
168	28224	12.9615	.00595238	208	43264	14.4222	.00480769
169	28561	13.0000	.00591716	209	43681	14.4568	.00478469
170	28900	13.0384	.00588235	210	44100	14.4914	.00476190
171	29241	13.0767	.00584795	211	44521	14.5258	.00473934
172	29584	13.1149	.00581395	212	44944	14.5602	.00471698
173	29929	13.1529	.00578035	213	45369	14.5945	.00469484
174	30276	13.1909	.00574713	214	45796	14.6287	.00467290
175	30625	13.2288	.00571429	215	46225	14.6629	.00465116
176	30976	13.2665	.00568182	216	46656	14.6969	.00462963
177	31329	13.3041	.00564972	217	47089	14.7309	.00460829
178	31684	13.3417	.00561798	218	47524	14.7648	.00458716
179	32041	13.3791	.00558659	219	47961	14.7986	.00456621
180	32400	13.4164	.00555556	220	48400	14.8324	.00454545
181	32761	13.4536	.00552486	221	48841	14.8661	.00452489
182	33124	13.4907	.00549451	222	49284	14.8997	.00450450
183	33489	13.5277	.00546448	223	49729	14.9332	.00448430
184	33856	13.5647	.00543478	224	50176	14.9666	.00446429
185	34225	13.6015	.00540541	225	50625	15.0000	.00444444
186	34596	13.6382	.00537634	226	51076	15.0333	.00442478
187	34969	13.6748	.00534759	227	51529	15.0665	.00440529
188	35344	13.7113	.00531915	228	51984	15.0997	.00438596
189	35721	13.7477	.00529101	229	52441	15.1327	.00436681
190	36100	13.7840	.00526316	230	52900	15.1658	.00434783
191	36481	13.8203	.00523560	231	53361	15.1987	.00432900
192	36864	13.8564	.00520833	232	53824	15.2315	.00431034
193	37249	13.8924	.00518135	233	54289	15.2643	.00429185
194	37636	13.9284	.00515464	234	54756	15.2971	.00427350
195	38025	13.9642	.00512821	235	55225	15.3297	.00425532
196	38416	14.0000	.00510204	236	55696	15.3623	.00423729
197	38809	14.0357	.00507614	237	56169	15.3948	.00421941
198	39204	14.0712	.00505051	238	56644	15.4272	.00420168
199	39601	14.1067	.00502513	239	57121	15.4596	.00418410
200	40000	14.1421	.00500000	240	57600	15.4919	.00416667

* Portions of Table II have been reproduced from J. W. Dunlap and A. K. Kurtz. *Handbook of Statistical Nomographs, Tables, and Formulas,* World Book Company, New York (1932), by permission of the authors and publishers.

TABLE II. *Table of Squares, Square Roots, and Reciprocals of Numbers from 1 to 1000*—Continued*

N	N^2	$\sqrt{N}$	$1/N$	N	N^2	$\sqrt{N}$	$1/N$
241	58081	15.5242	.00414938	281	78961	16.7631	.00355872
242	58564	15.5563	.00413223	282	79524	16.7929	.00354610
243	59049	15.5885	.00411523	283	80089	16.8226	.00353357
244	59536	15.6205	.00409836	284	80656	16.8523	.00352113
245	60025	15.6525	.00408163	285	81225	16.8819	.00350877
246	60516	15.6844	.00406504	286	81796	16.9115	.00349650
247	61009	15.7162	.00404858	287	82369	16.9411	.00348432
248	61504	15.7480	.00403226	288	82944	16.9706	.00347222
249	62001	15.7797	.00401606	289	83521	17.0000	.00346021
250	62500	15.8114	.00400000	290	84100	17 0294	.00344828
251	63001	15.8430	.00398406	291	84681	17.0587	.00343643
252	63504	15.8745	.00396825	292	85264	17.0880	.00342466
253	64009	15.9060	.00395257	293	85849	17.1172	.00341297
254	64516	15.9374	.00393701	294	86436	17.1464	.00340136
255	65025	15.9687	.00392157	295	87025	17.1756	.00338983
256	65536	16.0000	.00390625	296	87616	17.2047	.00337838
257	66049	16.0312	.00389105	297	88209	17.2337	.00336700
258	66564	16.0624	.00387597	298	88804	17.2627	.00335570
259	67081	16.0935	.00386100	299	89401	17.2916	.00334448
260	67600	16.1245	.00384615	300	90000	17.3205	.00333333
261	68121	16.1555	.00383142	301	90601	17.3494	.00332226
262	68644	16.1864	.00381679	302	91204	17.3781	.00331126
263	69169	16.2173	.00380228	303	91809	17.4069	.00330033
264	69696	16.2481	.00378788	304	92416	17.4356	.00328947
265	70225	16.2788	.00377358	305	93025	17.4642	.00327869
266	70756	16.3095	.00375940	306	93636	17.4929	.00326797
267	71289	16.3401	.00374532	307	94249	17.5214	.00325733
268	71824	16.3707	.00373134	308	94864	17.5499	.00324675
269	72361	16.4012	.00371747	309	95481	17.5784	.00323625
270	72900	16.4317	.00370370	310	96100	17.6068	.00322581
271	73441	16.4621	.00369004	311	96721	17.6352	.00321543
272	73984	16.4924	.00367647	312	97344	17.6635	.00320513
273	74529	16.5227	.00366300	313	97969	17.6918	.00319489
274	75076	16.5529	.00364964	314	98596	17.7200	.00318471
275	75625	16.5831	.00363636	315	99225	17.7482	.00317460
276	76176	16.6132	.00362319	316	99856	17.7764	.00316456
277	76729	16.6433	.00361011	317	100489	17.8045	.00315457
278	77284	16.6733	.00359712	318	101124	17.8326	.00314465
279	77841	16.7033	.00358423	319	101761	17.8606	.00313480
280	78400	16.7332	.00357143	320	102400	17.8885	.00312500

* Portions of Table II have been reproduced from J. W. Dunlap and A. K. Kurtz. *Handbook of Statistical Nomographs, Tables, and Formulas,* World Book Company, New York (1932), by permission of the authors and publishers.

TABLE II. *Table of Squares, Square Roots, and Reciprocals of Numbers from 1 to 1000*—Continued*

N	N²	√N	1/N	N	N²	√N	1/N
321	103041	17.9165	.00311526	361	130321	19.0000	.00277008
322	103684	17.9444	.00310559	362	131044	19.0263	.00276243
323	104329	17.9722	.00309598	363	131769	19.0526	.00275482
324	104976	18.0000	.00308642	364	132496	19.0788	.00274725
325	105625	18.0278	.00307692	365	133225	19.1050	.00273973
326	106276	18.0555	.00306748	366	133956	19.1311	.00273224
327	106929	18.0831	.00305810	367	134689	19.1572	.00272480
328	107584	18.1108	.00304878	368	135424	19.1833	.00271739
329	108241	18.1384	.00303951	369	136161	19.2094	.00271003
330	108900	18.1659	.00303030	370	136900	19.2354	.00270270
331	109561	18.1934	.00302115	371	137641	19.2614	.00269542
332	110224	18.2209	.00301205	372	138384	19.2873	.00268817
333	110889	18.2483	.00300300	373	139129	19.3132	.00268097
334	111556	18.2757	.00299401	374	139876	19.3391	.00267380
335	112225	18.3030	.00298507	375	140625	19.3649	.00266667
336	112896	18.3303	.00297619	376	141376	19.3907	.00265957
337	113569	18.3576	.00296736	377	142129	19.4165	.00265252
338	114244	18.3848	.00295858	378	142884	19.4422	.00264550
339	114921	18.4120	.00294985	379	143641	19.4679	.00263852
340	115600	18.4391	.00294118	380	144400	19.4936	.00263158
341	116281	18.4662	.00293255	381	145161	19.5192	.00262467
342	116964	18.4932	.00292398	382	145924	19.5448	.00261780
343	117649	18.5203	.00291545	383	146689	19.5704	.00261097
344	118336	18.5472	.00290698	384	147456	19.5959	.00260417
345	119025	18.5742	.00289855	385	148225	19.6214	.00259740
346	119716	18.6011	.00289017	386	148996	19.6469	.00259067
347	120409	18.6279	.00288184	387	149769	19.6723	.00258398
348	121104	18.6548	.00287356	388	150544	19.6977	.00257732
349	121801	18.6815	.00286533	389	151321	19.7231	.00257069
350	122500	18.7083	.00285714	390	152100	19.7484	.00256410
351	123201	18.7350	.00284900	391	152881	19.7737	.00255754
352	123904	18.7617	.00284091	392	153664	19.7990	.00255102
353	124609	18.7883	.00283286	393	154449	19.8242	.00254453
354	125316	18.8149	.00282486	394	155236	19.8494	.00253807
355	126025	18.8414	.00281690	395	156025	19.8746	.00253165
356	126736	18.8680	.00280899	396	156816	19.8997	.00252525
357	127449	18.8944	.00280112	397	157609	19.9249	.00251889
358	128164	18.9209	.00279330	398	158404	19.9499	.00251256
359	128881	18.9473	.00278552	399	159201	19.9750	.00250627
360	129600	18.9737	.00277778	400	160000	20.0000	.00250000

• Portions of Table II have been reproduced from J. W. Dunlap and A. K. Kurtz. *Handbook of Statistical Nomographs, Tables, and Formulas,* World Book Company, New York (1932), by permission of the authors and publishers.

TABLE II. *Table of Squares, Square Roots, and Reciprocals of Numbers from 1 to 1000*—Continued*

N	N^2	$\sqrt{N}$	$1/N$	N	N^2	$\sqrt{N}$	$1/N$
401	160801	20.0250	.00249377	441	194481	21.0000	.00226757
402	161604	20.0499	.00248756	442	195364	21.0238	.00226244
403	162409	20.0749	.00248139	443	196249	21.0476	.00225734
404	163216	20.0998	.00247525	444	197136	21.0713	.00225225
405	164025	20.1246	.00246914	445	198025	21.0950	.00224719
406	164836	20.1494	.00246305	446	198916	21.1187	.00224215
407	165649	20.1742	.00245700	447	199809	21.1424	.00223714
408	166464	20.1990	.00245098	448	200704	21.1660	.00223214
409	167281	20.2237	.00244499	449	201601	21.1896	.00222717
410	168100	20.2485	.00243902	450	202500	21.2132	.00222222
411	168921	20.2731	.00243309	451	203401	21.2368	.00221729
412	169744	20.2978	.00242718	452	204304	21.2603	.00221239
413	170569	20.3224	.00242131	453	205209	21.2838	.00220751
414	171396	20.3470	.00241546	454	206116	21.3073	.00220264
415	172225	20.3715	.00240964	455	207025	21.3307	.00219780
416	173056	20.3961	.00240385	456	207936	21.3542	.00219298
417	173889	20.4206	.00239808	457	208849	21.3776	.00218818
418	174724	20.4450	.00239234	458	209764	21.4009	.00218341
419	175561	20.4695	.00238663	459	210681	21.4243	.00217865
420	176400	20.4939	.00238095	460	211600	21.4476	.00217391
421	177241	20.5183	.00237530	461	212521	21.4709	.00216920
422	178084	20.5426	.00236967	462	213444	21.4942	.00216450
423	178929	20.5670	.00236407	463	214369	21.5174	.00215983
424	179776	20.5913	.00235849	464	215296	21.5407	.00215517
425	180625	20.6155	.00235294	465	216225	21.5639	.00215054
426	181476	20.6398	.00234742	466	217156	21.5870	.00214592
427	182329	20.6640	.00234192	467	218089	21.6102	.00214133
428	183184	20.6882	.00233645	468	219024	21.6333	.00213675
429	184041	20.7123	.00233100	469	219961	21.6564	.00213220
430	184900	20.7364	.00232558	470	220900	21.6795	.00212766
431	185761	20.7605	.00232019	471	221841	21.7025	.00212314
432	186624	20.7846	.00231481	472	222784	21.7256	.00211864
433	187489	20.8087	.00230947	473	223729	21.7486	.00211416
434	188356	20.8327	.00230415	474	224676	21.7715	.00210970
435	189225	20.8567	.00229885	475	225625	21.7945	.00210526
436	190096	20.8806	.00229358	476	226576	21.8174	.00210084
437	190969	20.9045	.00228833	477	227529	21.8403	.00209644
438	191844	20.9284	.00228311	478	228484	21.8632	.00209205
439	192721	20.9523	.00227790	479	229441	21.8861	.00208768
440	193600	20.9762	.00227273	480	230400	21.9089	.00208333

* Portions of Table II have been reproduced from J. W. Dunlap and A. K. Kurtz *Handbook of Statistical Nomographs, Tables, and Formulas,* World Book Company, New York (1932), by permission of the authors and publishers.

TABLE II. *Table of Squares, Square Roots, and Reciprocals of Numbers from 1 to 1000*—Continued*

N	N^2	$\sqrt{N}$	$1/N$	N	N^2	$\sqrt{N}$	$1/N$
481	231361	21.9317	.00207900	521	271441	22.8254	.00191939
482	232324	21.9545	.00207469	522	272484	22.8473	.00191571
483	233289	21.9773	.00207039	523	273529	22.8692	.00191205
484	234256	22.0000	.00206612	524	274576	22.8910	.00190840
485	235225	22.0227	.00206186	525	275625	22.9129	.00190476
486	236196	22.0454	.00205761	526	276676	22.9347	.00190114
487	237169	22.0681	.00205339	527	277729	22.9565	.00189753
488	238144	22.0907	.00204918	528	278784	22.9783	.00189394
489	239121	22.1133	.00204499	529	279841	23.0000	.00189036
490	240100	22.1359	.00204082	530	280900	23.0217	.00188679
491	241081	22.1585	.00203666	531	281961	23.0434	.00188324
492	242064	22.1811	.00203252	532	283024	23.0651	.00187970
493	243049	22.2036	.00202840	533	284089	23.0868	.00187617
494	244036	22.2261	.00202429	534	285156	23.1084	.00187266
495	245025	22.2486	.00202020	535	286225	23.1301	.00186916
496	246016	22.2711	.00201613	536	287296	23.1517	.00186567
497	247009	22.2935	.00201207	537	288369	23.1733	.00186220
498	248004	22.3159	.00200803	538	289444	23.1948	.00185874
499	249001	22.3383	.00200401	539	290521	23.2164	.00185529
500	250000	22.3607	.00200000	540	291600	23.2379	.00185185
501	251001	22.3830	.00199601	541	292681	23.2594	.00184843
502	252004	22.4054	.00199203	542	293764	23.2809	.00184502
503	253009	22.4277	.00198807	543	294849	23.3024	.00184162
504	254016	22.4499	.00198413	544	295936	23.3238	.00183824
505	255025	22.4722	.00198020	545	297025	23.3452	.00183486
506	256036	22.4944	.00197628	546	298116	23.3666	.00183150
507	257049	22.5167	.00197239	547	299209	23.3880	.00182815
508	258064	22.5389	.00196850	548	300304	23.4094	.00182482
509	259081	22.5610	.00196464	549	301401	23.4307	.00182149
510	260100	22.5832	.00196078	550	302500	23.4521	.00181818
511	261121	22.6053	.00195695	551	303601	23.4734	.00181488
512	262144	22.6274	.00195312	552	304704	23.4947	.00181159
513	263169	22.6495	.00194932	553	305809	23.5160	.00180832
514	264196	22.6716	.00194553	554	306916	23.5372	.00180505
515	265225	22.6936	.00194175	555	308025	23.5584	.00180180
516	266256	22.7156	.00193798	556	309136	23.5797	.00179856
517	267289	22.7376	.00193424	557	310249	23.6008	.00179533
518	268324	22.7596	.00193050	558	311364	23.6220	.00179211
519	269361	22.7816	.00192678	559	312481	23.6432	.00178891
520	270400	22.8035	.00192308	560	313600	23.6643	.00178571

* Portions of Table II have been reproduced from J. W. Dunlap and A. K. Kurtz. *Handbook of Statistical Nomographs, Tables, and Formulas,* World Book Company, New York (1932), by permission of the authors and publishers.

TABLE II. *Table of Squares, Square Roots, and Reciprocals of Numbers from 1 to 1000*—Continued*

N	N²	$\sqrt{N}$	1/N	N	N²	$\sqrt{N}$	1/N
561	314721	23.6854	.00178253	601	361201	24.5153	.00166389
562	315844	23.7065	.00177936	602	362404	24.5357	.00166113
563	316969	23.7276	.00177620	603	363609	24.5561	.00165837
564	318096	23.7487	.00177305	604	364816	24.5764	.00165563
565	319225	23.7697	.00176991	605	366025	24.5967	.00165289
566	320356	23.7908	.00176678	606	367236	24.6171	.00165017
567	321489	23.8118	.00176367	607	368449	24.6374	.00164745
568	322624	23.8328	.00176056	608	369664	24.6577	.00164474
569	323761	23.8537	.00175747	609	370881	24.6779	.00164204
570	324900	23.8747	.00175439	610	372100	24.6982	.00163934
571	326041	23.8956	.00175131	611	373321	24.7184	.00163666
572	327184	23.9165	.00174825	612	374544	24.7386	.00163399
573	328329	23.9374	.00174520	613	375769	24.7588	.00163132
574	329476	23.9583	.00174216	614	376996	24.7790	.00162866
575	330625	23.9792	.00173913	615	378225	24.7992	.00162602
576	331776	24.0000	.00173611	616	379456	24.8193	.00162338
577	332929	24.0208	.00173310	617	380689	24.8395	.00162075
578	334084	24.0416	.00173010	618	381924	24.8596	.00161812
579	335241	24.0624	.00172712	619	383161	24.8797	.00161551
580	336400	24.0832	.00172414	620	384400	24.8998	.00161290
581	337561	24.1039	.00172117	621	385641	24.9199	.00161031
582	338724	24.1247	.00171821	622	386884	24.9399	.00160772
583	339889	24.1454	.00171527	623	388129	24.9600	.00160514
584	341056	24.1661	.00171233	624	389376	24.9800	.00160256
585	342225	24.1868	.00170940	625	390625	25.0000	.00160000
586	343396	24.2074	.00170648	626	391876	25.0200	.00159744
587	344569	24.2281	.00170358	627	393129	25.0400	.00159490
588	345744	24.2487	.00170068	628	394384	25.0599	.00159236
589	346921	24.2693	.00169779	629	395641	25.0799	.00158983
590	348100	24.2899	.00169492	630	396900	25.0998	.00158730
591	349281	24.3105	.00169205	631	398161	25.1197	.00158479
592	350464	24.3311	.00168919	632	399424	25.1396	.00158228
593	351649	24.3516	.00168634	633	400689	25.1595	.00157978
594	352836	24.3721	.00168350	634	401956	25.1794	.00157729
595	354025	24.3926	.00168067	635	403225	25.1992	.00157480
596	355216	24.4131	.00167785	636	404496	25.2190	.00157233
597	356409	24.4336	.00167504	637	405769	25.2389	.00156986
598	357604	24.4540	.00167224	638	407044	25.2587	.00156740
599	358801	24.4745	.00166945	639	408321	25.2784	.00156495
600	360000	24.4949	.00166667	640	409600	25.2982	.00156250

* Portions of Table II have been reproduced from J. W. Dunlap and A. K. Kurtz. *Handbook of Statistical Nomographs, Tables, and Formulas,* World Book Company, New York (1932), by permission of the authors and publishers.

TABLE II. *Table of Squares, Square Roots, and Reciprocals of Numbers from 1 to 1000—Continued*

N	N^2	$\sqrt{N}$	$1/N$	N	N^2	$\sqrt{N}$	$1/N$
641	410881	25.3180	.00156006	681	463761	26.0960	.00146843
642	412164	25.3377	.00155763	682	465124	26.1151	.00146628
643	413449	25.3574	.00155521	683	466489	26.1343	.00146413
644	414736	25.3772	.00155280	684	467856	26.1534	.00146199
645	416025	25.3969	.00155039	685	469225	26.1725	.00145985
646	417316	25.4165	.00154799	686	470596	26.1916	.00145773
647	418609	25.4362	.00154560	687	471969	26.2107	.00145560
648	419904	25.4558	.00154321	688	473344	26.2298	.00145349
649	421201	25.4755	.00154083	689	474721	26.2488	.00145138
650	422500	25.4951	.00153846	690	476100	26.2679	.00144928
651	423801	25.5147	.00153610	691	477481	26.2869	.00144718
652	425104	25.5343	.00153374	692	478864	26.3059	.00144509
653	426409	25.5539	.00153139	693	480249	26.3249	.00144300
654	427716	25.5734	.00152905	694	481636	26.3439	.00144092
655	429025	25.5930	.00152672	695	483025	26.3629	.00143885
656	430336	25.6125	.00152439	696	484416	26.3818	.00143678
657	431649	25.6320	.00152207	697	485809	26.4008	.00143472
658	432964	25.6515	.00151976	698	487204	26.4197	.00143266
659	434281	25.6710	.00151745	699	488601	26.4386	.00143062
660	435600	25.6905	.00151515	700	490000	26.4575	.00142857
661	436921	25.7099	.00151286	701	491401	26.4764	.00142653
662	438244	25.7294	.00151057	702	492804	26.4953	.00142450
663	439569	25.7488	.00150830	703	494209	26.5141	.00142248
664	440896	25.7682	.00150602	704	495616	26.5330	.00142045
665	442225	25.7876	.00150376	705	497025	26.5518	.00141844
666	443556	25.8070	.00150150	706	498436	26.5707	.00141643
667	444889	25.8263	.00149925	707	499849	26.5895	.00141443
668	446224	25.8457	.00149701	708	501264	26.6083	.00141243
669	447561	25.8650	.00149477	709	502681	26.6271	.00141044
670	448900	25.8844	.00149254	710	504100	26.6458	.00140845
671	450241	25.9037	.00149031	711	505521	26.6646	.00140647
672	451584	25.9230	.00148810	712	506944	26.6833	.00140449
673	452929	25.9422	.00148588	713	508369	26.7021	.00140252
674	454276	25.9615	.00148368	714	509796	26.7208	.00140056
675	455625	25.9808	.00148148	715	511225	26.7395	.00139860
676	456976	26.0000	.00147929	716	512656	26.7582	.00139665
677	458329	26.0192	.00147710	717	514089	26.7769	.00139470
678	459684	26.0384	.00147493	718	515524	26.7955	.00139276
679	461041	26.0576	.00147275	719	516961	26.8142	.00139082
680	462400	26.0768	.00147059	720	518400	26.8328	.00138889

* Portions of Table II have been reproduced from J. W. Dunlap and A. K. Kurtz. *Handbook of Statistical Nomographs, Tables, and Formulas*, World Book Company, New York (1932), by permission of the authors and publishers.

TABLE II. *Table of Squares, Square Roots, and Reciprocals of Numbers from 1 to 1000*—Continued*

N	N^2	$\sqrt{N}$	$1/N$	N	N^2	$\sqrt{N}$	$1/N$
721	519841	26.8514	.00138696	761	579121	27.5862	.00131406
722	521284	26.8701	.00138504	762	580644	27.6043	.00131234
723	522729	26.8887	.00138313	763	582169	27.6225	.00131062
724	524176	26.9072	.00138122	764	583696	27.6405	.00130890
725	525625	26.9258	.00137931	765	585225	27.6586	.00130719
726	527076	26.9444	.00137741	766	586756	27.6767	.00130548
727	528529	26.9629	.00137552	767	588289	27.6948	.00130378
728	529984	26.9815	.00137363	768	589824	27.7128	.00130208
729	531441	27.0000	.00137174	769	591361	27.7308	.00130039
730	532900	27.0185	.00136986	770	592900	27.7489	.00129870
731	534361	27.0370	.00136799	771	594441	27.7669	.00129702
732	535824	27.0555	.00136612	772	595984	27.7849	.00129534
733	537289	27.0740	.00136426	773	597529	27.8029	.00129366
734	538756	27.0924	.00136240	774	599076	27.8209	.00129199
735	540225	27.1109	.00136054	775	600625	27.8388	.00129032
736	541696	27.1293	.00135870	776	602176	27.8568	.00128866
737	543169	27.1477	.00135685	777	603729	27.8747	.00128700
738	544644	27.1662	.00135501	778	605284	27.8927	.00128535
739	546121	27.1846	.00135318	779	606841	27.9106	.00128370
740	547600	27.2029	.00135135	780	608400	27.9285	.00128205
741	549081	27.2213	.00134953	781	609961	27.9464	.00128041
742	550564	27.2397	.00134771	782	611524	27.9643	.00127877
743	552049	27.2580	.00134590	783	613089	27.9821	.00127714
744	553536	27.2764	.00134409	784	614656	28.0000	.00127551
745	555025	27.2947	.00134228	785	616225	28.0179	.00127389
746	556516	27.3130	.00134048	786	617796	28.0357	.00127226
747	558009	27.3313	.00133869	787	619369	28.0535	.00127065
748	559504	27.3496	.00133690	788	620944	28.0713	.00126904
749	561001	27.3679	.00133511	789	622521	28.0891	.00126743
750	562500	27.3861	.00133333	790	624100	28.1069	.00126582
751	564001	27.4044	.00133156	791	625681	28.1247	.00126422
752	565504	27.4226	.00132979	792	627264	28.1425	.00126263
753	567009	27.4408	.00132802	793	628849	28.1603	.00126103
754	568516	27.4591	.00132626	794	630436	28.1780	.00125945
755	570025	27.4773	.00132450	795	632025	28.1957	.00125786
756	571536	27.4955	.00132275	796	633616	28.2135	.00125628
757	573049	27.5136	.00132100	797	635209	28.2312	.00125471
758	574564	27.5318	.00131926	798	636804	28.2489	.00125313
759	576081	27.5500	.00131752	799	638401	28.2666	.00125156
760	577600	27.5681	.00131579	800	640000	28.2843	.00125000

* Portions of Table II have been reproduced from J. W. Dunlap and A. K. Kurtz. *Handbook of Statistical Nomographs, Tables, and Formulas,* World Book Company, New York (1932), by permission of the authors and publishers.

TABLE II. *Table of Squares, Square Roots, and Reciprocals of Numbers from 1 to 1000*—Continued*

N	N^2	$\sqrt{N}$	$1/N$	N	N^2	$\sqrt{N}$	$1/N$
801	641601	28.3019	.00124844	841	707281	29.0000	.00118906
802	643204	28.3196	.00124688	842	708964	29.0172	.00118765
803	644809	28.3373	.00124533	843	710649	29.0345	.00118624
804	646416	28.3549	.00124378	844	712336	29.0517	.00118483
805	648025	28.3725	.00124224	845	714025	29.0689	.00118343
806	649636	28.3901	.00124069	846	715716	29.0861	.00118203
807	651249	28.4077	.00123916	847	717409	29.1033	.00118064
808	652864	28.4253	.00123762	848	719104	29.1204	.00117925
809	654481	28.4429	.00123609	849	720801	29.1376	.00117786
810	656100	28.4605	.00123457	850	722500	29.1548	.00117647
811	657721	28.4781	.00123305	851	724201	29.1719	.00117509
812	659344	28.4956	.00123153	852	725904	29.1890	.00117371
813	660969	28.5132	.00123001	853	727609	29.2062	.00117233
814	662596	28.5307	.00122850	854	729316	29.2233	.00117096
815	664225	28.5482	.00122699	855	731025	29.2404	.00116959
816	665856	28.5657	.00122549	856	732736	29.2575	.00116822
817	667489	28.5832	.00122399	857	734449	29.2746	.00116686
818	669124	28.6007	.00122249	858	736164	29.2916	.00116550
819	670761	28.6182	.00122100	859	737881	29.3087	.00116414
820	672400	28.6356	.00121951	860	739600	29.3258	.00116279
821	674041	28.6531	.00121803	861	741321	29.3428	.00116144
822	675684	28.6705	.00121655	862	743044	29.3598	.00116009
823	677329	28.6880	.00121507	863	744769	29.3769	.00115875
824	678976	28.7054	.00121359	864	746496	29.3939	.00115741
825	680625	28.7228	.00121212	865	748225	29.4109	.00115607
826	682276	28.7402	.00121065	866	749956	29.4279	.00115473
827	683929	28.7576	.00120919	867	751689	29.4449	.00115340
828	685584	28.7750	.00120773	868	753424	29.4618	.00115207
829	687241	28.7924	.00120627	869	755161	29.4788	.00115075
830	688900	28.8097	.00120482	870	756900	29.4958	.00114943
831	690561	28.8271	.00120337	871	758641	29.5127	.00114811
832	692224	28.8444	.00120192	872	760384	29.5296	.00114679
833	693889	28.8617	.00120048	873	762129	29.5466	.00114548
834	695556	28.8791	.00119904	874	763876	29.5635	.00114416
835	697225	28.8964	.00119760	875	765625	29.5804	.00114286
836	698896	28.9137	.00119617	876	767376	29.5973	.00114155
837	700569	28.9310	.00119474	877	769129	29.6142	.00114025
838	702244	28.9482	.00119332	878	770884	29.6311	.00113895
839	703921	28.9655	.00119190	879	772641	29.6479	.00113766
840	705600	28.9828	.00119048	880	774400	29.6648	.00113636

* Portions of Table II have been reproduced from J. W. Dunlap and A. K. Kurtz. *Handbook of Statistical Nomographs, Tables, and Formulas*, World Book Company, New York (1932), by permission of the authors and publishers.

TABLE II. *Table of Squares, Square Roots, and Reciprocals of Numbers from 1 to 1000—Continued*

N	N^2	$\sqrt{N}$	$1/N$	N	N^2	$\sqrt{N}$	$1/N$
881	776161	29.6816	.00113507	921	848241	30.3480	.00108578
882	777924	29.6985	.00113379	922	850084	30.3645	.00108460
883	779689	29.7153	.00113250	923	851929	30.3809	.00108342
884	781456	29.7321	.00113122	924	853776	30.3974	.00108225
885	783225	29.7489	.00112994	925	855625	30.4138	.00108108
886	784996	29.7658	.00112867	926	857476	30.4302	.00107991
887	786769	29.7825	.00112740	927	859329	30.4467	.00107875
888	788544	29.7993	.00112613	928	861184	30.4631	.00107759
889	790321	29.8161	.00112486	929	863041	30.4795	.00107643
890	792100	29.8329	.00112360	930	864900	30.4959	.00107527
891	793881	29.8496	.00112233	931	866761	30.5123	.00107411
892	795664	29.8664	.00112108	932	868624	30.5287	.00107296
893	797449	29.8831	.00111982	933	870489	30.5450	.00107181
894	799236	29.8998	.00111857	934	872356	30.5614	.00107066
895	801025	29.9166	.00111732	935	874225	30.5778	.00106952
896	802816	29.9333	.00111607	936	876096	30.5941	.00106838
897	804609	29.9500	.00111483	937	877969	30.6105	.00106724
898	806404	29.9666	.00111359	938	879844	30.6268	.00106610
899	808201	29.9833	.00111235	939	881721	30.6431	.00106496
900	810000	30.0000	.00111111	940	883600	30.6594	.00106383
901	811801	30.0167	.00110988	941	885481	30.6757	.00106270
902	813604	30.0333	.00110865	942	887364	30.6920	.00106157
903	815409	30.0500	.00110742	943	889249	30.7083	.00106045
904	817216	30.0666	.00110619	944	891136	30.7246	.00105932
905	819025	30.0832	.00110497	945	893025	30.7409	.00105820
906	820836	30.0998	.00110375	946	894916	30.7571	.00105708
907	822649	30.1164	.00110254	947	896809	30.7734	.00105597
908	824464	30.1330	.00110132	948	898704	30.7896	.00105485
909	826281	30.1496	.00110011	949	900601	30.8058	.00105374
910	828100	30.1662	.00109890	950	902500	30.8221	.00105263
911	829921	30.1828	.00109769	951	904401	30.8383	.00105152
912	831744	30.1993	.00109649	952	906304	30.8545	.00105042
913	833569	30.2159	.00109529	953	908209	30.8707	.00104932
914	835396	30.2324	.00109409	954	910116	30.8869	.00104822
915	837225	30.2490	.00109290	955	912025	30.9031	.00104712
916	839056	30.2655	.00109170	956	913936	30.9192	.00104603
917	840889	30.2820	.00109051	957	915849	30.9354	.00104493
918	842724	30.2985	.00108932	958	917764	30.9516	.00104384
919	844561	30.3150	.00108814	959	919681	30.9677	.00104275
920	846400	30.3315	.00108696	960	921600	30.9839	.00104167

*Portions of Table II have been reproduced from J. W. Dunlap and A. K. Kurtz. *Handbook of Statistical Nomographs, Tables, and Formulas,* World Book Company, New York (1932), by permission of the authors and publishers.

TABLE II. *Table of Squares, Square Roots, and Reciprocals
of Numbers from 1 to 1000*—Concluded*

N	N^2	$\sqrt{N}$	$1/N$	N	N^2	$\sqrt{N}$	$1/N$
961	923521	31.0000	.00104058	981	962361	31.3209	.00101937
962	925444	31.0161	.00103950	982	964324	31.3369	.00101833
963	927369	31.0322	.00103842	983	966289	31.3528	.00101729
964	929296	31.0483	.00103734	984	968256	31.3688	.00101626
965	931225	31.0644	.00103627	985	970225	31.3847	.00101523
966	933156	31.0805	.00103520	986	972196	31.4006	.00101420
967	935089	31.0966	.00103413	987	974169	31.4166	.00101317
968	937024	31.1127	.00103306	988	976144	31.4325	.00101215
969	938961	31.1288	.00103199	989	978121	31.4484	.00101112
970	940900	31.1448	.00103093	990	980100	31.4643	.00101010
971	942841	31.1609	.00102987	991	982081	31.4802	.00100908
972	944784	31.1769	.00102881	992	984064	31.4960	.00100806
973	946729	31.1929	.00102775	993	986049	31.5119	.00100705
974	948676	31.2090	.00102669	994	988036	31.5278	.00100604
975	950625	31.2250	.00102564	995	990025	31.5436	.00100503
976	952576	31.2410	.00102459	996	992016	31.5595	.00100402
977	954529	31.2570	.00102354	997	994009	31.5753	.00100301
978	956484	31.2730	.00102249	998	996004	31.5911	.00100200
979	958441	31.2890	.00102145	999	998001	31.6070	.00100100
980	960400	31.3050	.00102041	1000	1000000	31.6228	.00100000

* Portions of Table II have been reproduced from J. W. Dunlap and A. K. Kurtz,
Handbook of Statistical Nomographs, Tables, and Formulas, World Book Company, New York
(1932), by permission of the authors and publishers.

TABLE III. *Areas and Ordinates of the Standard Normal Curve*

(1) Z	(2) A AREA FROM μ TO Z	(3) B AREA IN LARGER PORTION	(4) C AREA IN SMALLER PORTION	(5) y ORDINATE AT Z
0.00	.0000	.5000	.5000	.3989
0.01	.0040	.5040	.4960	.3989
0.02	.0080	.5080	.4920	.3989
0.03	.0120	.5120	.4880	.3988
0.04	.0160	.5160	.4840	.3986
0.05	.0199	.5199	.4801	.3984
0.06	.0239	.5239	.4761	.3982
0.07	.0279	.5279	.4721	.3980
0.08	.0319	.5319	.4681	.3977
0.09	.0359	.5359	.4641	.3973
0.10	.0398	.5398	.4602	.3970
0.11	.0438	.5438	.4562	.3965
0.12	.0478	.5478	.4522	.3961
0.13	.0517	.5517	.4483	.3956
0.14	.0557	.5557	.4443	.3951
0.15	.0596	.5596	.4404	.3945
0.16	.0636	.5636	.4364	.3939
0.17	.0675	.5675	.4325	.3932
0.18	.0714	.5714	.4286	.3925
0.19	.0753	.5753	.4247	.3918
0.20	.0793	.5793	.4207	.3910
0.21	.0832	.5832	.4168	.3902
0.22	.0871	.5871	.4129	.3894
0.23	.0910	.5910	.4090	.3885
0.24	.0948	.5948	.4052	.3876
0.25	.0987	.5987	.4013	.3867
0.26	.1026	.6026	.3974	.3857
0.27	.1064	.6064	.3936	.3847
0.28	.1103	.6103	.3897	.3836
0.29	.1141	.6141	.3859	.3825
0.30	.1179	.6179	.3821	.3814
0.31	.1217	.6217	.3783	.3802
0.32	.1255	.6255	.3745	.3790
0.33	.1293	.6293	.3707	.3778
0.34	.1331	.6331	.3669	.3765

TABLE III. *Areas and Ordinates of the Standard Normal Curve—Continued*

(1) Z	(2) A AREA FROM μ TO Z	(3) B AREA IN LARGER PORTION	(4) C AREA IN SMALLER PORTION	(5) y ORDINATE AT Z
0.35	.1368	.6368	.3632	.3752
0.36	.1406	.6406	.3594	.3739
0.37	.1443	.6443	.3557	.3725
0.38	.1480	.6480	.3520	.3712
0.39	.1517	.6517	.3483	.3697
0.40	.1554	.6554	.3446	.3683
0.41	.1591	.6591	.3409	.3668
0.42	.1628	.6628	.3372	.3653
0.43	.1664	.6664	.3336	.3637
0.44	.1700	.6700	.3300	.3621
				.3605
0.45	.1736	.6736	.3264	
0.46	.1772	.6772	.3228	.3589
0.47	.1808	.6808	.3192	.3572
0.48	.1844	.6844	.3156	.3555
0.49	.1879	.6879	.3121	.3538
0.50	.1915	.6915	.3085	.3521
0.51	.1950	.6950	.3050	.3503
0.52	.1985	.6985	.3015	.3485
0.53	.2019	.7019	.2981	.3467
0.54	.2054	.7054	.2946	.3448
0.55	.2088	.7088	.2912	.3429
0.56	.2123	.7123	.2877	.3410
0.57	.2157	.7157	.2843	.3391
0.58	.2190	.7190	.2810	.3372
0.59	.2224	.7224	.2776	.3352
0.60	.2257	.7257	.2743	.3332
0.61	.2291	.7291	.2709	.3312
0.62	.2324	.7324	.2676	.3292
0.63	.2357	.7357	.2643	.3271
0.64	.2389	.7389	.2611	.3251
0.65	.2422	.7422	.2578	.3230
0.66	.2454	.7454	.2546	.3209
0.67	.2486	.7486	.2514	.3187
0.68	.2517	.7517	.2483	.3166
0.69	.2549	.7549	.2451	.3144

TABLE III. *Areas and Ordinates of the Standard Normal Curve—Continued*

(1) Z	(2) A AREA FROM μ TO Z	(3) B AREA IN LARGER PORTION	(4) C AREA IN SMALLER PORTION	(5) y ORDINATE AT Z
0.70	.2580	.7580	.2420	.3123
0.71	.2611	.7611	.2389	.3101
0.72	.2642	.7642	.2358	.3079
0.73	.2673	.7673	.2327	.3056
0.74	.2704	.7704	.2296	.3034
0.75	.2734	.7734	.2266	.3011
0.76	.2764	.7764	.2236	.2989
0.77	.2794	.7794	.2206	.2966
0.78	.2823	.7823	.2177	.2943
0.79	.2852	.7852	.2148	.2920
0.80	.2881	.7881	.2119	.2897
0.81	.2910	.7910	.2090	.2874
0.82	.2939	.7939	.2061	.2850
0.83	.2967	.7967	.2033	.2827
0.84	.2995	.7995	.2005	.2803
0.85	.3023	.8023	.1977	.2780
0.86	.3051	.8051	.1949	.2756
0.87	.3078	.8078	.1922	.2732
0.88	.3106	.8106	.1894	.2709
0.89	.3133	.8133	.1867	.2685
0.90	.3159	.8159	.1841	.2661
0.91	.3186	.8186	.1814	.2637
0.92	.3212	.8212	.1788	.2613
0.93	.3238	.8238	.1762	.2589
0.94	.3264	.8264	.1736	.2565
0.95	.3289	.8289	.1711	.2541
0.96	.3315	.8315	.1685	.2516
0.97	.3340	.8340	.1660	.2492
0.98	.3365	.8365	.1635	.2468
0.99	.3389	.8389	.1611	.2444
1.00	.3413	.8413	.1587	.2420
1.01	.3438	.8438	.1562	.2396
1.02	.3461	.8461	.1539	.2371
1.03	.3485	.8485	.1515	.2347
1.04	.3508	.8508	.1492	.2323

TABLE III. *Areas and Ordinates of the Standard Normal Curve—Continued*

(1) Z	(2) A AREA FROM μ TO Z	(3) B AREA IN LARGER PORTION	(4) C AREA IN SMALLER PORTION	(5) y ORDINATE AT Z
1.05	.3531	.8531	.1469	.2299
1.06	.3554	.8554	.1446	.2275
1.07	.3577	.8577	.1423	.2251
1.08	.3599	.8599	.1401	.2227
1.09	.3621	.8621	.1379	.2203
1.10	.3643	.8643	.1357	.2179
1.11	.3665	.8665	.1335	.2155
1.12	.3686	.8686	.1314	.2131
1.13	.3708	.8708	.1292	.2107
1.14	.3729	.8729	.1271	.2083
1.15	.3749	.8749	.1251	.2059
1.16	.3770	.8770	.1230	.2036
1.17	.3790	.8790	.1210	.2012
1.18	.3810	.8810	.1190	.1989
1.19	.3830	.8830	.1170	.1965
1.20	.3849	.8849	.1151	.1942
1.21	.3869	.8869	.1131	.1919
1.22	.3888	.8888	.1112	.1895
1.23	.3907	.8907	.1093	.1872
1.24	.3925	.8925	.1075	.1849
1.25	.3944	.8944	.1056	.1826
1.26	.3962	.8962	.1038	.1804
1.27	.3980	.8980	.1020	.1781
1.28	.3997	.8997	.1003	.1758
1.29	.4015	.9015	.0985	.1736
1.30	.4032	.9032	.0968	.1714
1.31	.4049	.9049	.0951	.1691
1.32	.4066	.9066	.0934	.1669
1.33	.4082	.9082	.0918	.1647
1.34	.4099	.9099	.0901	.1626
1.35	.4115	.9115	.0885	.1604
1.36	.4131	.9131	.0869	.1582
1.37	.4147	.9147	.0853	.1561
1.38	.4162	.9162	.0838	.1539
1.39	.4177	.9177	.0823	.1518

TABLE III. *Areas and Ordinates of the Standard Normal Curve—Continued*

(1) Z	(2) A AREA FROM μ TO Z	(3) B AREA IN LARGER PORTION	(4) C AREA IN SMALLER PORTION	(5) y ORDINATE AT Z
1.40	.4192	.9192	.0808	.1497
1.41	.4207	.9207	.0793	.1476
1.42	.4222	.9222	.0778	.1456
1.43	.4236	.9236	.0764	.1435
1.44	.4251	.9251	.0749	.1415
1.45	.4265	.9265	.0735	.1394
1.46	.4279	.9279	.0721	.1374
1.47	.4292	.9292	.0708	.1354
1.48	.4306	.9306	.0694	.1334
1.49	.4319	.9319	.0681	.1315
1.50	.4332	.9332	.0668	.1295
1.51	.4345	.9345	.0655	.1276
1.52	.4357	.9357	.0643	.1257
1.53	.4370	.9370	.0630	.1238
1.54	.4382	.9382	.0618	.1219
1.55	.4394	.9394	.0606	.1200
1.56	.4406	.9406	.0594	.1182
1.57	.4418	.9418	.0582	.1163
1.58	.4429	.9429	.0571	.1145
1.59	.4441	.9441	.0559	.1127
1.60	.4452	.9452	.0548	.1109
1.61	.4463	.9463	.0537	.1092
1.62	.4474	.9474	.0526	.1074
1.63	.4484	.9484	.0516	.1057
1.64	.4495	.9495	.0505	.1040
1.65	.4505	.9505	.0495	.1023
1.66	.4515	.9515	.0485	.1006
1.67	.4525	.9525	.0475	.0989
1.68	.4535	.9535	.0465	.0973
1.69	.4545	.9545	.0455	.0957
1.70	.4554	.9554	.0446	.0940
1.71	.4564	.9564	.0436	.0925
1.72	.4573	.9573	.0427	.0909
1.73	.4582	.9582	.0418	.0893
1.74	.4591	.9591	.0409	.0878

TABLE III. *Areas and Ordinates of the Standard Normal Curve—Continued*

(1) Z	(2) A AREA FROM μ TO Z	(3) B AREA IN LARGER PORTION	(4) C AREA IN SMALLER PORTION	(5) y ORDINATE AT Z
1.75	.4599	.9599	.0401	.0863
1.76	.4608	.9608	.0392	.0848
1.77	.4616	.9616	.0384	.0833
1.78	.4625	.9625	.0375	.0818
1.79	.4633	.9633	.0367	.0804
1.80	.4641	.9641	.0359	.0790
1.81	.4649	.9649	.0351	.0775
1.82	.4656	.9656	.0344	.0761
1.83	.4664	.9664	.0336	.0748
1.84	.4671	.9671	.0329	.0734
1.85	.4678	.9678	.0322	.0721
1.86	.4686	.9686	.0314	.0707
1.87	.4693	.9693	.0307	.0694
1.88	.4699	.9699	.0301	.0681
1.89	.4706	.9706	.0294	.0669
1.90	.4713	.9713	.0287	.0656
1.91	.4719	.9719	.0281	.0644
1.92	.4726	.9726	.0274	.0632
1.93	.4732	.9732	.0268	.0620
1.94	.4738	.9738	.0262	.0608
1.95	.4744	.9744	.0256	.0596
1.96	.4750	.9750	.0250	.0584
1.97	.4756	.9756	.0244	.0573
1.98	.4761	.9761	.0239	.0562
1.99	.4767	.9767	.0233	.0551
2.00	.4772	.9772	.0228	.0540
2.01	.4778	.9778	.0222	.0529
2.02	.4783	.9783	.0217	.0519
2.03	.4788	.9788	.0212	.0508
2.04	.4793	.9793	.0207	.0498
2.05	.4798	.9798	.0202	.0488
2.06	.4803	.9803	.0197	.0478
2.07	.4808	.9808	.0192	.0468
2.08	.4812	.9812	.0188	.0459
2.09	.4817	.9817	.0183	.0449

TABLE III. *Areas and Ordinates of the Standard Normal Curve—Continued*

(1) Z	(2) A Area from μ to Z	(3) B Area in Larger Portion	(4) C Area in Smaller Portion	(5) y Ordinate at Z
2.10	.4821	.9821	.0179	.0440
2.11	.4826	.9826	.0174	.0431
2.12	.4830	.9830	.0170	.0422
2.13	.4834	.9834	.0166	.0413
2.14	.4838	.9838	.0162	.0404
2.15	.4842	.9842	.0158	.0396
2.16	.4846	.9846	.0154	.0387
2.17	.4850	.9850	.0150	.0379
2.18	.4854	.9854	.0146	.0371
2.19	.4857	.9857	.0143	.0363
2.20	.4861	.9861	.0139	.0355
2.21	.4864	.9864	.0136	.0347
2.22	.4868	.9868	.0132	.0339
2.23	.4871	.9871	.0129	.0332
2.24	.4875	.9875	.0125	.0325
2.25	.4878	.9878	.0122	.0317
2.26	.4881	.9881	.0119	.0310
2.27	.4884	.9884	.0116	.0303
2.28	.4887	.9887	.0113	.0297
2.29	.4890	.9890	.0110	.0290
2.30	.4893	.9893	.0107	.0283
2.31	.4896	.9896	.0104	.0277
2.32	.4898	.9898	.0102	.0270
2.33	.4901	.9901	.0099	.0264
2.34	.4904	.9904	.0096	.0258
2.35	.4906	.9906	.0094	.0252
2.36	.4909	.9909	.0091	.0246
2.37	.4911	.9911	.0089	.0241
2.38	.4913	.9913	.0087	.0235
2.39	.4916	.9916	.0084	.0229
2.40	.4918	.9918	.0082	.0224
2.41	.4920	.9920	.0080	.0219
2.42	.4922	.9922	.0078	.0213
2.43	.4925	.9925	.0075	.0208
2.44	.4927	.9927	.0073	.0203

TABLE III. *Areas and Ordinates of the Standard Normal Curve—Continued*

(1) Z	(2) A AREA FROM μ TO Z	(3) B AREA IN LARGER PORTION	(4) C AREA IN SMALLER PORTION	(5) y ORDINATE AT Z
2.45	.4929	.9929	.0071	.0198
2.46	.4931	.9931	.0069	.0194
2.47	.4932	.9932	.0068	.0189
2.48	.4934	.9934	.0066	.0184
2.49	.4936	.9936	.0064	.0180
2.50	.4938	.9938	.0062	.0175
2.51	.4940	.9940	.0060	.0171
2.52	.4941	.9941	.0059	.0167
2.53	.4943	.9943	.0057	.0163
2.54	.4945	.9945	.0055	.0158
2.55	.4946	.9946	.0054	.0154
2.56	.4948	.9948	.0052	.0151
2.57	.4949	.9949	.0051	.0147
2.58	.4951	.9951	.0049	.0143
2.59	.4952	.9952	.0048	.0139
2.60	.4953	.9953	.0047	.0136
2.61	.4955	.9955	.0045	.0132
2.62	.4956	.9956	.0044	.0129
2.63	.4957	.9957	.0043	.0126
2.64	.4959	.9959	.0041	.0122
2.65	.4960	.9960	.0040	.0119
2.66	.4961	.9961	.0039	.0116
2.67	.4962	.9962	.0038	.0113
2.68	.4963	.9963	.0037	.0110
2.69	.4964	.9964	.0036	.0107
2.70	.4965	.9965	.0035	.0104
2.71	.4966	.9966	.0034	.0101
2.72	.4967	.9967	.0033	.0099
2.73	.4968	.9968	.0032	.0096
2.74	.4969	.9969	.0031	.0093
2.75	.4970	.9970	.0030	.0091
2.76	.4971	.9971	.0029	.0088
2.77	.4972	.9972	.0028	.0086
2.78	.4973	.9973	.0027	.0084
2.79	.4974	.9974	.0026	.0081

TABLE III. *Areas and Ordinates of the Standard Normal Curve—Continued*

(1) Z	(2) A AREA FROM μ TO Z	(3) B AREA IN LARGER PORTION	(4) C AREA IN SMALLER PORTION	(5) y ORDINATE AT Z
2.80	.4974	.9974	.0026	.0079
2.81	.4975	.9975	.0025	.0077
2.82	.4976	.9976	.0024	.0075
2.83	.4977	.9977	.0023	.0073
2.84	.4977	.9977	.0023	.0071
2.85	.4978	.9978	.0022	.0069
2.86	.4979	.9979	.0021	.0067
2.87	.4979	.9979	.0021	.0065
2.88	.4980	.9980	.0020	.0063
2.89	.4981	.9981	.0019	.0061
2.90	.4981	.9981	.0019	.0060
2.91	.4982	.9982	.0018	.0058
2.92	.4982	.9982	.0018	.0056
2.93	.4983	.9983	.0017	.0055
2.94	.4984	.9984	.0016	.0053
2.95	.4984	.9984	.0016	.0051
2.96	.4985	.9985	.0015	.0050
2.97	.4985	.9985	.0015	.0048
2.98	.4986	.9986	.0014	.0047
2.99	.4986	.9986	.0014	.0046
3.00	.4987	.9987	.0013	.0044
3.01	.4987	.9987	.0013	.0043
3.02	.4987	.9987	.0013	.0042
3.03	.4988	.9988	.0012	.0040
3.04	.4988	.9988	.0012	.0039
3.05	.4989	.9989	.0011	.0038
3.06	.4989	.9989	.0011	.0037
3.07	.4989	.9989	.0011	.0036
3.08	.4990	.9990	.0010	.0035
3.09	.4990	.9990	.0010	.0034
3.10	.4990	.9990	.0010	.0033
3.11	.4991	.9991	.0009	.0032
3.12	.4991	.9991	.0009	.0031
3.13	.4991	.9991	.0009	.0030
3.14	.4992	.9992	.0008	.0029

TABLE III. *Areas and Ordinates of the Standard Normal Curve—Concluded*

(1) Z	(2) A AREA FROM μ TO Z	(3) B AREA IN LARGER PORTION	(4) C AREA IN SMALLER PORTION	(5) y ORDINATE AT Z
3.15	.4992	.9992	.0008	.0028
3.16	.4992	.9992	.0008	.0027
3.17	.4992	.9992	.0008	.0026
3.18	.4993	.9993	.0007	.0025
3.19	.4993	.9993	.0007	.0025
3.20	.4993	.9993	.0007	.0024
3.21	.4993	.9993	.0007	.0023
3.22	.4994	.9994	.0006	.0022
3.23	.4994	.9994	.0006	.0022
3.24	.4994	.9994	.0006	.0021
3.30	.4995	.9995	.0005	.0017
3.40	.4997	.9997	.0003	.0012
3.50	.4998	.9998	.0002	.0009
3.60	.4998	.9998	.0002	.0006
3.70	.4999	.9999	.0001	.0004

TABLE IV. *Table of* t*

The probabilities given by the column headings are for a one-sided test, assuming a null hypothesis to be true. For example, with 20 d.f., we have $P(t \geq 2.845) = 0.005$. Similarly, we have $P(t \leq -2.845) = 0.005$. For a two-sided test, we have $P(t \geq 2.845) + P(t \leq -2.845) = 0.005 + 0.005 = 0.01$.

df \ P	.25	.10	.05	.025	.01	.005	.0025	.001
1	1.000	3.078	6.314	12.706	31,821	63.657	127.321	318.309
2	.816	1.886	2.920	4.303	6.965	9.925	14.089	22.327
3	.765	1.638	2.353	3.182	4.541	5.841	7.453	10.214
4	.741	1.533	2.132	2.776	3.747	4.604	5.598	7.173
5	.727	1.476	2.015	2.571	3.365	4.032	4.773	5.893
6	.718	1.440	1.943	2.447	3.143	3.707	4.317	5.208
7	.711	1.415	1.895	2.365	2.998	3.499	4.029	4.785
8	.706	1.397	1.860	2.306	2.896	3.355	3.833	4.501
9	.703	1.383	1.833	2.262	2.821	3.250	3.690	4.297
10	.700	1.372	1.812	2.228	2.764	3.169	3.581	4.144
11	.697	1.363	1.796	2.201	2.718	3.106	3.497	4.025
12	.695	1.356	1.782	2.179	2.681	3.055	3.428	3.930
13	.694	1.350	1.771	2.160	2.650	3.012	3.372	3.852
14	.692	1.345	1.761	2.145	2.624	2.977	3.326	3.787
15	.691	1.341	1.753	2.131	2.602	2.947	3.286	3.733
16	.690	1.337	1.746	2.120	2.583	2.921	3.252	3.686
17	.689	1.333	1.740	2.110	2.567	2.898	3.223	3.646
18	.688	1.330	1.734	2.101	2.552	2.878	3.197	3.610
19	.688	1.328	1.729	2.093	2.539	2.861	3.174	3.579
20	.687	1.325	1.725	2.086	2.528	2.845	3.153	3.552
21	.686	1.323	1.721	2.080	2.518	2.831	3.135	3.527
22	.686	1.321	1.717	2.074	2.508	2.819	3.119	3.505
23	.685	1.319	1.714	2.069	2.500	2.807	3.104	3.485
24	.685	1.318	1.711	2.064	2.492	2.797	3.090	3.467
25	.684	1.316	1.708	2.060	2.485	2.787	3.078	3.450
26	.684	1.315	1.706	2.056	2.479	2.779	3.067	3.435
27	.684	1.314	1.703	2.052	2.473	2.771	3.057	3.421
28	.683	1.313	1.701	2.048	2.467	2.763	3.047	3.408
29	.683	1.311	1.699	2.045	2.462	2.756	3.038	3.396
30	.683	1.310	1.697	2.042	2.457	2.750	3.030	3.385
35	.682	1.306	1.690	2.030	2.438	2.724	2.996	3.340
40	.681	1.303	1.684	2.021	2.423	2.704	2.971	3.307
45	.680	1.301	1.679	2.014	2.412	2.690	2.952	3.281
50	.679	1.299	1.676	2.009	2.403	2.678	2.937	3.261
55	.679	1.297	1.673	2.004	2.396	2.668	2.925	3.245
60	.679	1.296	1.671	2.000	2.390	2.660	2.915	3.232
70	.678	1.294	1.667	1.994	2.381	2.648	2.899	3.211
80	.678	1.292	1.664	1.990	2.374	2.639	2.887	3.195
90	.677	1.291	1.662	1.987	2.368	2.632	2.878	3.183
100	.677	1.290	1.660	1.984	2.364	2.626	2.871	3.174
200	.676	1.286	1.652	1.972	2.345	2.601	2.838	3.131
500	.675	1.283	1.648	1.965	2.334	2.586	2.820	3.107
1,000	.675	1.282	1.646	1.962	2.330	2.581	2.813	3.098
2,000	.675	1.282	1.645	1.961	2.328	2.578	2.810	3.094
10,000	.675	1.282	1.645	1.960	2.327	2.576	2.808	3.091
∞	.674	1.282	1.645	1.960	2.326	2.576	2.807	3.090

* Table IV is reprinted from Enrico T. Federighi, Extended Tables of the Percentage Points of Student's *t* Distribution. *Journal of the American Statistical Association*, 1959, Volume 54, Number 267, pp. 683-688, by permission.

TABLE IV. *Table of* t*—*Continued*

df \ P	.0005	.00025	.0001	.00005	.000025	.00001
1	636.619	1,273.239	3,183.099	6,366.198	12,732.395	31,830.989
2	31.598	44.705	70.700	99.992	141.416	223.603
3	12.924	16.326	22.204	28.000	35.298	47.928
4	8.610	10.306	13.034	15.544	18.522	23.332
5	6.869	7.976	9.678	11.178	12.893	15.547
6	5.959	6.788	8.025	9.082	10.261	12.032
7	5.408	6.082	7.063	7.885	8.782	10.103
8	5.041	5.618	6.442	7.120	7.851	8.907
9	4.781	5.291	6.010	6.594	7.215	8.102
10	4.587	5.049	5.694	6.211	6.757	7.527
11	4.437	4.863	5.453	5.921	6.412	7.098
12	4.318	4.716	5.263	5.694	6.143	6.756
13	4.221	4.597	5.111	5.513	5.928	6.501
14	4.140	4.499	4.985	5.363	5.753	6.287
15	4.073	4.417	4.880	5.239	5.607	6.109
16	4.015	4.346	4.791	5.134	5.484	5.960
17	3.965	4.286	4.714	5.044	5.379	5.832
18	3.922	4.233	4.648	4.966	5.288	5.722
19	3.883	4.187	4.590	4.897	5.209	5.627
20	3.850	4.146	4.539	4.837	5.139	5.543
21	3.819	4.110	4.493	4.784	5.077	5.469
22	3.792	4.077	4.452	4.736	5.022	5.402
23	3.768	4.048	4.415	4.693	4.972	5.343
24	3.745	4.021	4.382	4.654	4.927	5.290
25	3.725	3.997	4.352	4.619	4.887	5.241
26	3.707	3.974	4.324	4.587	4.850	5.197
27	3.690	3.954	4.299	4.558	4.816	5.157
28	3.674	3.935	4.275	4.530	4.784	5.120
29	3.659	3.918	4.254	4.506	4.756	5.086
30	3.646	3.902	4.234	4.482	4.729	5.054
35	3.591	3.836	4.153	4.389	4.622	4.927
40	3.551	3.788	4.094	4.321	4.544	4.835
45	3.520	3.752	4.049	4.269	4.485	4.766
50	3.496	3.723	4.014	4.228	4.438	4.711
55	3.476	3.700	3.986	4.196	4.401	4.667
60	3.460	3.681	3.962	4.169	4.370	4.631
70	3.435	3.651	3.926	4.127	4.323	4.576
80	3.416	3.629	3.899	4.096	4.288	4.535
90	3.402	3.612	3.878	4.072	4.261	4.503
100	3.390	3.598	3.862	4.053	4.240	4.478
200	3.340	3.539	3.789	3.970	4.146	4.369
500	3.310	3.504	3.747	3.922	4.091	4.306
1,000	3.300	3.492	3.763	3.906	4.073	4.285
2,000	3.295	3.486	3.726	3.898	4.064	4.275
10,000	3.292	3.482	3.720	3.892	4.058	4.267
∞	3.291	3.481	3.719	3.891	4.056	4.265

* Table IV is reprinted from Enrico T. Federighi, Extended Tables of the Percentage Points of Student's *t* Distribution. *Journal of the American Statistical Association,* 1959, Volume 54, Number 267, pp. 683-688, by permission.

TABLE IV. *Table of t*—Continued*

df	.000005	.0000025	.000001	.0000005	.00000025	.0000001
1	63,661.977	127,323.954	318,309.886	636,619.772	1,273,239.545	3,183,098.862
2	316.225	447.212	707.106	999,999	1,414.213	2,236.068
3	60.397	76.104	103.299	130.155	163.989	222.572
4	27.771	33.047	41.578	49.459	58.829	73.986
5	17.897	20.591	24.771	28.477	32.734	39.340
6	13.555	15.260	17.830	20.047	22.532	26.286
7	11.215	12.437	14.241	15.764	17.447	19.932
8	9.782	10.731	12.110	13.257	14.504	16.320
9	8.827	9.605	10.720	11.637	12.623	14.041
10	8.150	8.812	9.752	10.516	11.328	12.492
11	7.648	8.227	9.043	9.702	10.397	11.381
12	7.261	7.780	8.504	9.085	9.695	10.551
13	6.955	7.427	8.082	8.604	9.149	9.909
14	6.706	7.142	7.743	8.218	8.713	9.400
15	6.502	6.907	7.465	7.903	8.358	8.986
16	6.330	6.711	7.233	7.642	8.064	8.645
17	6.184	6.545	7.037	7.421	7.817	8.358
18	6.059	6.402	6.869	7.232	7.605	8.115
19	5.949	6.278	6.723	7.069	7.423	7.905
20	5.854	6.170	6.597	6.927	7.265	7.723
21	5.769	6.074	6.485	6.802	7.126	7.564
22	5.694	5.989	6.386	6.692	7.003	7.423
23	5.627	5.913	6.297	6.593	6.893	7.298
24	5.566	5.845	6.218	6.504	6.795	7.185
25	5.511	5.783	6.146	6.424	6.706	7.085
26	5.461	5.726	6.081	6.352	6.626	6.993
27	5.415	5.675	6.021	6.286	6.553	6.910
28	5.373	5.628	5.967	6.225	6.486	6.835
29	5.335	5.585	5.917	6.170	6.426	6.765
30	5.299	5.545	5.871	6.119	6.369	6.701
35	5.156	5.385	5.687	5.915	6.143	6.447
40	5.053	5.269	5.554	5.768	5.983	6.266
45	4.975	5.182	5.454	5.659	5.862	6.130
50	4.914	5.115	5.377	5.573	5.769	6.025
55	4.865	5.060	5.315	5.505	5.694	5.942
60	4.825	5.015	5.264	5.449	5.633	5.873
70	4.763	4.946	5.185	5.363	5.539	5.768
80	4.717	4.896	5.128	5.300	5.470	5.691
90	4.682	4.857	5.084	5.252	5.417	5.633
100	4.654	4.826	5.049	5.214	5.376	5.587
200	4.533	4.692	4.897	5.048	5.196	5.387
500	4.463	4.615	4.810	4.953	5.094	5.273
1,000	4.440	4.590	4.781	4.922	5.060	5.236
2,000	4.428	4.578	4.767	4.907	5.043	5.218
10,000	4.419	4.567	4.756	4.895	5.029	5.203
∞	4.417	4.565	4.753	4.892	5.026	5.199

* Table IV is reprinted from Enrico T. Federighi. Extended Tables of the Percentage Points of Student's *t* Distribution. *Journal of the American Statistical Association,* 1959, Volume 54, Number 267, pp. 683-688, by permission.

TABLE V. *Table of* F*

The values of F significant at the 0.05 (Roman type) and 0.01 (boldface type) levels of significance with n_1 degrees of freedom for the numerator and n_2 degrees of freedom for the denominator of the F ratio.

n_1 degrees of freedom

n_2	1	2	3	4	5	6	7	8	9	10	11	12	14	16	20	24	30	40	50	75	100	200	500	∞
1	161 **4,052**	200 **4,999**	216 **5,403**	225 **5,625**	230 **5,764**	234 **5,859**	237 **5,928**	239 **5,981**	241 **6,022**	242 **6,056**	243 **6,082**	244 **6,106**	245 **6,142**	246 **6,169**	248 **6,208**	249 **6,234**	250 **6,258**	251 **6,286**	252 **6,302**	253 **6,323**	253 **6,334**	254 **6,352**	254 **6,361**	254 **6,366**
2	18.51 **98.49**	19.00 **99.00**	19.16 **99.17**	19.25 **99.25**	19.30 **99.30**	19.33 **99.33**	19.36 **99.34**	19.37 **99.36**	19.38 **99.38**	19.39 **99.40**	19.40 **99.41**	19.41 **99.42**	19.42 **99.43**	19.43 **99.44**	19.44 **99.45**	19.45 **99.46**	19.46 **99.47**	19.47 **99.48**	19.47 **99.48**	19.48 **99.49**	19.49 **99.49**	19.49 **99.49**	19.50 **99.50**	19.50 **99.50**
3	10.13 **34.12**	9.55 **30.82**	9.28 **29.46**	9.12 **28.71**	9.01 **28.24**	8.94 **27.91**	8.88 **27.67**	8.84 **27.49**	8.81 **27.34**	8.78 **27.23**	8.76 **27.13**	8.74 **27.05**	8.71 **26.92**	8.69 **26.83**	8.66 **26.69**	8.64 **26.60**	8.62 **26.50**	8.60 **26.41**	8.58 **26.35**	8.57 **26.27**	8.56 **26.23**	8.54 **26.18**	8.54 **26.14**	8.53 **26.12**
4	7.71 **21.20**	6.94 **18.00**	6.59 **16.69**	6.39 **15.98**	6.26 **15.52**	6.16 **15.21**	6.09 **14.98**	6.04 **14.80**	6.00 **14.66**	5.96 **14.54**	5.93 **14.45**	5.91 **14.37**	5.87 **14.24**	5.84 **14.15**	5.80 **14.02**	5.77 **13.93**	5.74 **13.83**	5.71 **13.74**	5.70 **13.69**	5.68 **13.61**	5.66 **13.57**	5.65 **13.52**	5.64 **13.48**	5.63 **13.46**
5	6.61 **16.26**	5.79 **13.27**	5.41 **12.06**	5.19 **11.39**	5.05 **10.97**	4.95 **10.67**	4.88 **10.45**	4.82 **10.27**	4.78 **10.15**	4.74 **10.05**	4.70 **9.96**	4.68 **9.89**	4.64 **9.77**	4.60 **9.68**	4.56 **9.55**	4.53 **9.47**	4.50 **9.38**	4.46 **9.29**	4.44 **9.24**	4.42 **9.17**	4.40 **9.13**	4.38 **9.07**	4.37 **9.04**	4.36 **9.02**
6	5.99 **13.74**	5.14 **10.92**	4.76 **9.78**	4.53 **9.15**	4.39 **8.75**	4.28 **8.47**	4.21 **8.26**	4.15 **8.10**	4.10 **7.98**	4.06 **7.87**	4.03 **7.79**	4.00 **7.72**	3.96 **7.60**	3.92 **7.52**	3.87 **7.39**	3.84 **7.31**	3.81 **7.23**	3.77 **7.14**	3.75 **7.09**	3.72 **7.02**	3.71 **6.99**	3.69 **6.94**	3.68 **6.90**	3.67 **6.88**
7	5.59 **12.25**	4.74 **9.55**	4.35 **8.45**	4.12 **7.85**	3.97 **7.46**	3.87 **7.19**	3.79 **7.00**	3.73 **6.84**	3.68 **6.71**	3.63 **6.62**	3.60 **6.54**	3.57 **6.47**	3.52 **6.35**	3.49 **6.27**	3.44 **6.15**	3.41 **6.07**	3.38 **5.98**	3.34 **5.90**	3.32 **5.85**	3.29 **5.78**	3.28 **5.75**	3.25 **5.70**	3.24 **5.67**	3.23 **5.65**
8	5.32 **11.26**	4.46 **8.65**	4.07 **7.59**	3.84 **7.01**	3.69 **6.63**	3.58 **6.37**	3.50 **6.19**	3.44 **6.03**	3.39 **5.91**	3.34 **5.82**	3.31 **5.74**	3.28 **5.67**	3.23 **5.56**	3.20 **5.48**	3.15 **5.36**	3.12 **5.28**	3.08 **5.20**	3.05 **5.11**	3.03 **5.06**	3.00 **5.00**	2.98 **4.96**	2.96 **4.91**	2.94 **4.88**	2.93 **4.86**
9	5.12 **10.56**	4.26 **8.02**	3.86 **6.99**	3.63 **6.42**	3.48 **6.06**	3.37 **5.80**	3.29 **5.62**	3.23 **5.47**	3.18 **5.35**	3.13 **5.26**	3.10 **5.18**	3.07 **5.11**	3.02 **5.00**	2.98 **4.92**	2.93 **4.80**	2.90 **4.73**	2.86 **4.64**	2.82 **4.56**	2.80 **4.51**	2.77 **4.45**	2.76 **4.41**	2.73 **4.36**	2.72 **4.33**	2.71 **4.31**
10	4.96 **10.04**	4.10 **7.56**	3.71 **6.55**	3.48 **5.99**	3.33 **5.64**	3.22 **5.39**	3.14 **5.21**	3.07 **5.06**	3.02 **4.95**	2.97 **4.85**	2.94 **4.78**	2.91 **4.71**	2.86 **4.60**	2.82 **4.52**	2.77 **4.41**	2.74 **4.33**	2.70 **4.25**	2.67 **4.17**	2.64 **4.12**	2.61 **4.05**	2.59 **4.01**	2.56 **3.96**	2.55 **3.93**	2.54 **3.91**
11	4.84 **9.65**	3.98 **7.20**	3.59 **6.22**	3.36 **5.67**	3.20 **5.32**	3.09 **5.07**	3.01 **4.88**	2.95 **4.74**	2.90 **4.63**	2.86 **4.54**	2.82 **4.46**	2.79 **4.40**	2.74 **4.29**	2.70 **4.21**	2.65 **4.10**	2.61 **4.02**	2.57 **3.94**	2.53 **3.86**	2.50 **3.80**	2.47 **3.74**	2.45 **3.70**	2.42 **3.66**	2.41 **3.62**	2.40 **3.60**
12	4.75 **9.33**	3.88 **6.93**	3.49 **5.95**	3.26 **5.41**	3.11 **5.06**	3.00 **4.82**	2.92 **4.65**	2.85 **4.50**	2.80 **4.39**	2.76 **4.30**	2.72 **4.22**	2.69 **4.16**	2.64 **4.05**	2.60 **3.98**	2.54 **3.86**	2.50 **3.78**	2.46 **3.70**	2.42 **3.61**	2.40 **3.56**	2.36 **3.49**	2.35 **3.46**	2.32 **3.41**	2.31 **3.38**	2.30 **3.36**
13	4.67 **9.07**	3.80 **6.70**	3.41 **5.74**	3.18 **5.20**	3.02 **4.86**	2.92 **4.62**	2.84 **4.44**	2.77 **4.30**	2.72 **4.19**	2.67 **4.10**	2.63 **4.02**	2.60 **3.96**	2.55 **3.85**	2.51 **3.78**	2.46 **3.67**	2.42 **3.59**	2.38 **3.51**	2.34 **3.42**	2.32 **3.37**	2.28 **3.30**	2.26 **3.27**	2.24 **3.21**	2.22 **3.18**	2.21 **3.16**

* Table V is reproduced from Snedecor: *Statistical Methods*, Iowa State College Press, Ames, Iowa, by permission of the author and publisher.

TABLE V. Table of F*—Continued

n_1 degrees of freedom

n_2	1	2	3	4	5	6	7	8	9	10	11	12	14	16	20	24	30	40	50	75	100	200	500	∞
14	4.60 / 8.86	3.74 / 6.51	3.34 / 5.56	3.11 / 5.03	2.96 / 4.69	2.85 / 4.46	2.77 / 4.28	2.70 / 4.14	2.65 / 4.03	2.60 / 3.94	2.56 / 3.86	2.53 / 3.80	2.48 / 3.70	2.44 / 3.62	2.39 / 3.51	2.35 / 3.43	2.31 / 3.34	2.27 / 3.26	2.24 / 3.21	2.21 / 3.14	2.19 / 3.11	2.16 / 3.06	2.14 / 3.02	2.13 / 3.00
15	4.54 / 8.68	3.68 / 6.36	3.29 / 5.42	3.06 / 4.89	2.90 / 4.56	2.79 / 4.32	2.70 / 4.14	2.64 / 4.00	2.59 / 3.89	2.55 / 3.80	2.51 / 3.73	2.48 / 3.67	2.43 / 3.56	2.39 / 3.48	2.33 / 3.36	2.29 / 3.29	2.25 / 3.20	2.21 / 3.12	2.18 / 3.07	2.15 / 3.00	2.12 / 2.97	2.10 / 2.92	2.08 / 2.89	2.07 / 2.87
16	4.49 / 8.53	3.63 / 6.23	3.24 / 5.29	3.01 / 4.77	2.85 / 4.44	2.74 / 4.20	2.66 / 4.03	2.59 / 3.89	2.54 / 3.78	2.49 / 3.69	2.45 / 3.61	2.42 / 3.55	2.37 / 3.45	2.33 / 3.37	2.28 / 3.25	2.24 / 3.18	2.20 / 3.10	2.16 / 3.01	2.13 / 2.96	2.09 / 2.89	2.07 / 2.86	2.04 / 2.80	2.02 / 2.77	2.01 / 2.75
17	4.45 / 8.40	3.59 / 6.11	3.20 / 5.18	2.96 / 4.67	2.81 / 4.34	2.70 / 4.10	2.62 / 3.93	2.55 / 3.79	2.50 / 3.68	2.45 / 3.59	2.41 / 3.52	2.38 / 3.45	2.33 / 3.35	2.29 / 3.27	2.23 / 3.16	2.19 / 3.08	2.15 / 3.00	2.11 / 2.92	2.08 / 2.86	2.04 / 2.79	2.02 / 2.76	1.99 / 2.70	1.97 / 2.67	1.96 / 2.65
18	4.41 / 8.28	3.55 / 6.01	3.16 / 5.09	2.93 / 4.58	2.77 / 4.25	2.66 / 4.01	2.58 / 3.85	2.51 / 3.71	2.46 / 3.60	2.41 / 3.51	2.37 / 3.44	2.34 / 3.37	2.29 / 3.27	2.25 / 3.19	2.19 / 3.07	2.15 / 3.00	2.11 / 2.91	2.07 / 2.83	2.04 / 2.78	2.00 / 2.71	1.98 / 2.68	1.95 / 2.62	1.93 / 2.59	1.92 / 2.57
19	4.38 / 8.18	3.52 / 5.93	3.13 / 5.01	2.90 / 4.50	2.74 / 4.17	2.63 / 3.94	2.55 / 3.77	2.48 / 3.63	2.43 / 3.52	2.38 / 3.43	2.34 / 3.36	2.31 / 3.30	2.26 / 3.19	2.21 / 3.12	2.15 / 3.00	2.11 / 2.92	2.07 / 2.84	2.02 / 2.76	2.00 / 2.70	1.96 / 2.63	1.94 / 2.60	1.91 / 2.54	1.90 / 2.51	1.88 / 2.49
20	4.35 / 8.10	3.49 / 5.85	3.10 / 4.94	2.87 / 4.43	2.71 / 4.10	2.60 / 3.87	2.52 / 3.71	2.45 / 3.56	2.40 / 3.45	2.35 / 3.37	2.31 / 3.30	2.28 / 3.23	2.23 / 3.13	2.18 / 3.05	2.12 / 2.94	2.08 / 2.86	2.04 / 2.77	1.99 / 2.69	1.96 / 2.63	1.92 / 2.56	1.90 / 2.53	1.87 / 2.47	1.85 / 2.44	1.84 / 2.42
21	4.32 / 8.02	3.47 / 5.78	3.07 / 4.87	2.84 / 4.37	2.68 / 4.04	2.57 / 3.81	2.49 / 3.65	2.42 / 3.51	2.37 / 3.40	2.32 / 3.31	2.28 / 3.24	2.25 / 3.17	2.20 / 3.07	2.15 / 2.99	2.09 / 2.88	2.05 / 2.80	2.00 / 2.72	1.96 / 2.63	1.93 / 2.58	1.89 / 2.51	1.87 / 2.47	1.84 / 2.42	1.82 / 2.38	1.81 / 2.36
22	4.30 / 7.94	3.44 / 5.72	3.05 / 4.82	2.82 / 4.31	2.66 / 3.99	2.55 / 3.76	2.47 / 3.59	2.40 / 3.45	2.35 / 3.35	2.30 / 3.26	2.26 / 3.18	2.23 / 3.12	2.18 / 3.02	2.13 / 2.94	2.07 / 2.83	2.03 / 2.75	1.98 / 2.67	1.93 / 2.58	1.91 / 2.53	1.87 / 2.46	1.84 / 2.42	1.81 / 2.37	1.80 / 2.33	1.78 / 2.31
23	4.28 / 7.88	3.42 / 5.66	3.03 / 4.76	2.80 / 4.26	2.64 / 3.94	2.53 / 3.71	2.45 / 3.54	2.38 / 3.41	2.32 / 3.30	2.28 / 3.21	2.24 / 3.14	2.20 / 3.07	2.14 / 2.97	2.10 / 2.89	2.04 / 2.78	2.00 / 2.70	1.96 / 2.62	1.91 / 2.53	1.88 / 2.48	1.84 / 2.41	1.82 / 2.37	1.79 / 2.32	1.77 / 2.28	1.76 / 2.26
24	4.26 / 7.82	3.40 / 5.61	3.01 / 4.72	2.78 / 4.22	2.62 / 3.90	2.51 / 3.67	2.43 / 3.50	2.36 / 3.36	2.30 / 3.25	2.26 / 3.17	2.22 / 3.09	2.18 / 3.03	2.13 / 2.93	2.09 / 2.85	2.02 / 2.74	1.98 / 2.66	1.94 / 2.58	1.89 / 2.49	1.86 / 2.44	1.82 / 2.36	1.80 / 2.33	1.76 / 2.27	1.74 / 2.23	1.73 / 2.21
25	4.24 / 7.77	3.38 / 5.57	2.99 / 4.68	2.76 / 4.18	2.60 / 3.86	2.49 / 3.63	2.41 / 3.46	2.34 / 3.32	2.28 / 3.21	2.24 / 3.13	2.20 / 3.05	2.16 / 2.99	2.11 / 2.89	2.06 / 2.81	2.00 / 2.70	1.96 / 2.62	1.92 / 2.54	1.87 / 2.45	1.84 / 2.40	1.80 / 2.32	1.77 / 2.29	1.74 / 2.23	1.72 / 2.19	1.71 / 2.17
26	4.22 / 7.72	3.37 / 5.53	2.98 / 4.64	2.74 / 4.14	2.59 / 3.82	2.47 / 3.59	2.39 / 3.42	2.32 / 3.29	2.27 / 3.17	2.22 / 3.09	2.18 / 3.02	2.15 / 2.96	2.10 / 2.86	2.05 / 2.77	1.99 / 2.66	1.95 / 2.58	1.90 / 2.50	1.85 / 2.41	1.82 / 2.36	1.78 / 2.28	1.76 / 2.25	1.72 / 2.19	1.70 / 2.15	1.69 / 2.13

* Table V is reproduced from Snedecor: Statistical Methods, Iowa State College Press, Ames, Iowa, by permission of the author and publisher.

TABLE V. *Table of F*—Continued

n_1 degrees of freedom

Each cell shows two values: the upper (5% point) and the lower (1% point, in bold).

n_2	1	2	3	4	5	6	7	8	9	10	11	12	14	16	20	24	30	40	50	75	100	200	500	∞
27	4.21 / 7.68	3.35 / 5.49	2.96 / 4.60	2.73 / 4.11	2.57 / 3.79	2.46 / 3.56	2.37 / 3.39	2.30 / 3.26	2.25 / 3.14	2.20 / 3.06	2.16 / 2.98	2.13 / 2.93	2.08 / 2.83	2.03 / 2.74	1.97 / 2.63	1.93 / 2.55	1.88 / 2.47	1.84 / 2.38	1.80 / 2.33	1.76 / 2.25	1.74 / 2.21	1.71 / 2.16	1.68 / 2.12	1.67 / 2.10
28	4.20 / 7.64	3.34 / 5.45	2.95 / 4.57	2.71 / 4.07	2.56 / 3.76	2.44 / 3.53	2.36 / 3.36	2.29 / 3.23	2.24 / 3.11	2.19 / 3.03	2.15 / 2.95	2.12 / 2.90	2.06 / 2.80	2.02 / 2.71	1.96 / 2.60	1.91 / 2.52	1.87 / 2.44	1.81 / 2.35	1.78 / 2.30	1.75 / 2.22	1.72 / 2.18	1.69 / 2.13	1.67 / 2.09	1.65 / 2.06
29	4.18 / 7.60	3.33 / 5.42	2.93 / 4.54	2.70 / 4.04	2.54 / 3.73	2.43 / 3.50	2.35 / 3.33	2.28 / 3.20	2.22 / 3.08	2.18 / 3.00	2.14 / 2.92	2.10 / 2.87	2.05 / 2.77	2.00 / 2.68	1.94 / 2.57	1.90 / 2.49	1.85 / 2.41	1.80 / 2.32	1.77 / 2.27	1.73 / 2.19	1.71 / 2.15	1.68 / 2.10	1.65 / 2.06	1.64 / 2.03
30	4.17 / 7.56	3.32 / 5.39	2.92 / 4.51	2.69 / 4.02	2.53 / 3.70	2.42 / 3.47	2.34 / 3.30	2.27 / 3.17	2.21 / 3.06	2.16 / 2.98	2.12 / 2.90	2.09 / 2.84	2.04 / 2.74	1.99 / 2.66	1.93 / 2.55	1.89 / 2.47	1.84 / 2.38	1.79 / 2.29	1.76 / 2.24	1.72 / 2.16	1.69 / 2.13	1.66 / 2.07	1.64 / 2.03	1.62 / 2.01
32	4.15 / 7.50	3.30 / 5.34	2.90 / 4.46	2.67 / 3.97	2.51 / 3.66	2.40 / 3.42	2.32 / 3.25	2.25 / 3.12	2.19 / 3.01	2.14 / 2.94	2.10 / 2.86	2.07 / 2.80	2.02 / 2.70	1.97 / 2.62	1.91 / 2.51	1.86 / 2.42	1.82 / 2.34	1.76 / 2.25	1.74 / 2.20	1.69 / 2.12	1.67 / 2.08	1.64 / 2.02	1.61 / 1.98	1.59 / 1.96
34	4.13 / 7.44	3.28 / 5.29	2.88 / 4.42	2.65 / 3.93	2.49 / 3.61	2.38 / 3.38	2.30 / 3.21	2.23 / 3.08	2.17 / 2.97	2.12 / 2.89	2.08 / 2.82	2.05 / 2.76	2.00 / 2.66	1.95 / 2.58	1.89 / 2.47	1.84 / 2.38	1.80 / 2.30	1.74 / 2.21	1.71 / 2.15	1.67 / 2.08	1.64 / 2.04	1.61 / 1.98	1.59 / 1.94	1.57 / 1.91
36	4.11 / 7.39	3.26 / 5.25	2.86 / 4.38	2.63 / 3.89	2.48 / 3.58	2.36 / 3.35	2.28 / 3.18	2.21 / 3.04	2.15 / 2.94	2.10 / 2.86	2.06 / 2.78	2.03 / 2.72	1.98 / 2.62	1.93 / 2.54	1.87 / 2.43	1.82 / 2.35	1.78 / 2.26	1.72 / 2.17	1.69 / 2.12	1.65 / 2.04	1.62 / 2.00	1.59 / 1.94	1.56 / 1.90	1.55 / 1.87
38	4.10 / 7.35	3.25 / 5.21	2.85 / 4.34	2.62 / 3.86	2.46 / 3.54	2.35 / 3.32	2.26 / 3.15	2.19 / 3.02	2.14 / 2.91	2.09 / 2.82	2.05 / 2.75	2.02 / 2.69	1.96 / 2.59	1.92 / 2.51	1.85 / 2.40	1.80 / 2.32	1.76 / 2.22	1.71 / 2.14	1.67 / 2.08	1.63 / 2.00	1.60 / 1.97	1.57 / 1.90	1.54 / 1.86	1.53 / 1.84
40	4.08 / 7.31	3.23 / 5.18	2.84 / 4.31	2.61 / 3.83	2.45 / 3.51	2.34 / 3.29	2.25 / 3.12	2.18 / 2.99	2.12 / 2.88	2.07 / 2.80	2.04 / 2.73	2.00 / 2.66	1.95 / 2.56	1.90 / 2.49	1.84 / 2.37	1.79 / 2.29	1.74 / 2.20	1.69 / 2.11	1.66 / 2.05	1.61 / 1.97	1.59 / 1.94	1.55 / 1.88	1.53 / 1.84	1.51 / 1.81
42	4.07 / 7.27	3.22 / 5.15	2.83 / 4.29	2.59 / 3.80	2.44 / 3.49	2.32 / 3.26	2.24 / 3.10	2.17 / 2.96	2.11 / 2.86	2.06 / 2.77	2.02 / 2.70	1.99 / 2.64	1.94 / 2.54	1.89 / 2.46	1.82 / 2.35	1.78 / 2.26	1.73 / 2.17	1.68 / 2.08	1.64 / 2.02	1.60 / 1.94	1.57 / 1.91	1.54 / 1.85	1.51 / 1.80	1.49 / 1.78
44	4.06 / 7.24	3.21 / 5.12	2.82 / 4.26	2.58 / 3.78	2.43 / 3.46	2.31 / 3.24	2.23 / 3.07	2.16 / 2.94	2.10 / 2.84	2.05 / 2.75	2.01 / 2.68	1.98 / 2.62	1.92 / 2.52	1.88 / 2.44	1.81 / 2.32	1.76 / 2.24	1.72 / 2.15	1.66 / 2.06	1.63 / 2.00	1.58 / 1.92	1.56 / 1.88	1.52 / 1.82	1.50 / 1.78	1.48 / 1.75
46	4.05 / 7.21	3.20 / 5.10	2.81 / 4.24	2.57 / 3.76	2.42 / 3.44	2.30 / 3.22	2.22 / 3.05	2.14 / 2.92	2.09 / 2.82	2.04 / 2.73	2.00 / 2.66	1.97 / 2.60	1.91 / 2.50	1.87 / 2.42	1.80 / 2.30	1.75 / 2.22	1.71 / 2.13	1.65 / 2.04	1.62 / 1.98	1.57 / 1.90	1.54 / 1.86	1.51 / 1.80	1.48 / 1.76	1.46 / 1.72
48	4.04 / 7.19	3.19 / 5.08	2.80 / 4.22	2.56 / 3.74	2.41 / 3.42	2.30 / 3.20	2.21 / 3.04	2.14 / 2.90	2.08 / 2.80	2.03 / 2.71	1.99 / 2.64	1.96 / 2.58	1.90 / 2.48	1.86 / 2.40	1.79 / 2.28	1.74 / 2.20	1.70 / 2.11	1.64 / 2.02	1.61 / 1.96	1.56 / 1.88	1.53 / 1.84	1.50 / 1.78	1.47 / 1.73	1.45 / 1.70

* Table V is reproduced by Snedecor: *Statistical Methods*, Iowa State College Press, Ames, Iowa, by permission of the author and publisher.

TABLE V. *Table of F*—*Continued*

n_1 degrees of freedom

n_2	1	2	3	4	5	6	7	8	9	10	11	12	14	16	20	24	30	40	50	75	100	200	500	∞
50	4.03 / 7.17	3.18 / 5.06	2.79 / 4.20	2.56 / 3.72	2.40 / 3.41	2.29 / 3.18	2.20 / 3.02	2.13 / 2.88	2.07 / 2.78	2.02 / 2.70	1.98 / 2.62	1.95 / 2.56	1.90 / 2.46	1.85 / 2.39	1.78 / 2.26	1.74 / 2.18	1.69 / 2.10	1.63 / 2.00	1.60 / 1.94	1.55 / 1.86	1.52 / 1.82	1.48 / 1.76	1.46 / 1.71	1.44 / 1.68
55	4.02 / 7.12	3.17 / 5.01	2.78 / 4.16	2.54 / 3.68	2.38 / 3.37	2.27 / 3.15	2.18 / 2.98	2.11 / 2.85	2.05 / 2.75	2.00 / 2.66	1.97 / 2.59	1.93 / 2.53	1.88 / 2.43	1.83 / 2.35	1.76 / 2.23	1.72 / 2.15	1.67 / 2.06	1.61 / 1.96	1.58 / 1.90	1.52 / 1.82	1.50 / 1.78	1.46 / 1.71	1.43 / 1.66	1.41 / 1.64
60	4.00 / 7.08	3.15 / 4.98	2.76 / 4.13	2.52 / 3.65	2.37 / 3.34	2.25 / 3.12	2.17 / 2.95	2.10 / 2.82	2.04 / 2.72	1.99 / 2.63	1.95 / 2.56	1.92 / 2.50	1.86 / 2.40	1.81 / 2.32	1.75 / 2.20	1.70 / 2.12	1.65 / 2.03	1.59 / 1.93	1.56 / 1.87	1.50 / 1.79	1.48 / 1.74	1.44 / 1.68	1.41 / 1.63	1.39 / 1.60
65	3.99 / 7.04	3.14 / 4.95	2.75 / 4.10	2.51 / 3.62	2.36 / 3.31	2.24 / 3.09	2.15 / 2.93	2.08 / 2.79	2.02 / 2.70	1.98 / 2.61	1.94 / 2.54	1.90 / 2.47	1.85 / 2.37	1.80 / 2.30	1.73 / 2.18	1.68 / 2.09	1.63 / 2.00	1.57 / 1.90	1.54 / 1.84	1.49 / 1.76	1.46 / 1.71	1.42 / 1.64	1.39 / 1.60	1.37 / 1.56
70	3.98 / 7.01	3.13 / 4.92	2.74 / 4.08	2.50 / 3.60	2.35 / 3.29	2.23 / 3.07	2.14 / 2.91	2.07 / 2.77	2.01 / 2.67	1.97 / 2.59	1.93 / 2.51	1.89 / 2.45	1.84 / 2.35	1.79 / 2.28	1.72 / 2.15	1.67 / 2.07	1.62 / 1.98	1.56 / 1.88	1.53 / 1.82	1.47 / 1.74	1.45 / 1.69	1.40 / 1.62	1.37 / 1.56	1.35 / 1.53
80	3.96 / 6.96	3.11 / 4.88	2.72 / 4.04	2.48 / 3.56	2.33 / 3.25	2.21 / 3.04	2.12 / 2.87	2.05 / 2.74	1.99 / 2.64	1.95 / 2.55	1.91 / 2.48	1.88 / 2.41	1.82 / 2.32	1.77 / 2.24	1.70 / 2.11	1.65 / 2.03	1.60 / 1.94	1.54 / 1.84	1.51 / 1.78	1.45 / 1.70	1.42 / 1.65	1.38 / 1.57	1.35 / 1.52	1.32 / 1.49
100	3.94 / 6.90	3.09 / 4.82	2.70 / 3.98	2.46 / 3.51	2.30 / 3.20	2.19 / 2.99	2.10 / 2.82	2.03 / 2.69	1.97 / 2.59	1.92 / 2.51	1.88 / 2.43	1.85 / 2.36	1.79 / 2.26	1.75 / 2.19	1.68 / 2.06	1.63 / 1.98	1.57 / 1.89	1.51 / 1.79	1.48 / 1.73	1.42 / 1.64	1.39 / 1.59	1.34 / 1.51	1.30 / 1.46	1.28 / 1.43
125	3.92 / 6.84	3.07 / 4.78	2.68 / 3.94	2.44 / 3.47	2.29 / 3.17	2.17 / 2.95	2.08 / 2.79	2.01 / 2.65	1.95 / 2.56	1.90 / 2.47	1.86 / 2.40	1.83 / 2.33	1.77 / 2.23	1.72 / 2.15	1.65 / 2.03	1.60 / 1.94	1.55 / 1.85	1.49 / 1.75	1.45 / 1.68	1.39 / 1.59	1.36 / 1.54	1.31 / 1.46	1.27 / 1.40	1.25 / 1.37
150	3.91 / 6.81	3.06 / 4.75	2.67 / 3.91	2.43 / 3.44	2.27 / 3.14	2.16 / 2.92	2.07 / 2.76	2.00 / 2.62	1.94 / 2.53	1.89 / 2.44	1.85 / 2.37	1.82 / 2.30	1.76 / 2.20	1.71 / 2.12	1.64 / 2.00	1.59 / 1.91	1.54 / 1.83	1.47 / 1.72	1.44 / 1.66	1.37 / 1.56	1.34 / 1.51	1.29 / 1.43	1.25 / 1.37	1.22 / 1.33
200	3.89 / 6.76	3.04 / 4.71	2.65 / 3.88	2.41 / 3.41	2.26 / 3.11	2.14 / 2.90	2.05 / 2.73	1.98 / 2.60	1.92 / 2.50	1.87 / 2.41	1.83 / 2.34	1.80 / 2.28	1.74 / 2.17	1.69 / 2.09	1.62 / 1.97	1.57 / 1.88	1.52 / 1.79	1.45 / 1.69	1.42 / 1.62	1.35 / 1.53	1.32 / 1.48	1.26 / 1.39	1.22 / 1.33	1.19 / 1.28
400	3.86 / 6.70	3.02 / 4.66	2.62 / 3.83	2.39 / 3.36	2.23 / 3.06	2.12 / 2.85	2.03 / 2.69	1.96 / 2.55	1.90 / 2.46	1.85 / 2.37	1.81 / 2.29	1.78 / 2.23	1.72 / 2.12	1.67 / 2.04	1.60 / 1.92	1.54 / 1.84	1.49 / 1.74	1.42 / 1.64	1.38 / 1.57	1.32 / 1.47	1.28 / 1.42	1.22 / 1.32	1.16 / 1.24	1.13 / 1.19
1000	3.85 / 6.66	3.00 / 4.62	2.61 / 3.80	2.38 / 3.34	2.22 / 3.04	2.10 / 2.82	2.02 / 2.66	1.95 / 2.53	1.89 / 2.43	1.84 / 2.34	1.80 / 2.26	1.76 / 2.20	1.70 / 2.09	1.65 / 2.01	1.58 / 1.89	1.53 / 1.81	1.47 / 1.71	1.41 / 1.61	1.36 / 1.54	1.30 / 1.44	1.26 / 1.38	1.19 / 1.28	1.13 / 1.19	1.08 / 1.11
∞	3.84 / 6.64	2.99 / 4.60	2.60 / 3.78	2.37 / 3.32	2.21 / 3.02	2.09 / 2.80	2.01 / 2.64	1.94 / 2.51	1.88 / 2.41	1.83 / 2.32	1.79 / 2.24	1.75 / 2.18	1.69 / 2.07	1.64 / 1.99	1.57 / 1.87	1.52 / 1.79	1.46 / 1.69	1.40 / 1.59	1.35 / 1.52	1.28 / 1.41	1.24 / 1.36	1.17 / 1.25	1.11 / 1.15	1.00 / 1.00

* Table V is reproduced from Snedecor: *Statistical Methods*, Iowa State College Press, Ames, Iowa, by permission of the author and publisher.

TABLE VI. *Table of* χ^2*

The probabilities given by the column headings are for obtaining χ^2 equal to or greater than the tabled values when a null hypothesis is true and when χ^2 has the degrees of freedom given by the column at the left.

DEGREES OF FREEDOM	P = .99	.98	.95	.90	.80	.70	.50	.30	.20	.10	.05	.02	.01
1	.000157	.000628	.00393	.0158	.0642	.148	.455	1.074	1.642	2.706	3.841	5.412	6.635
2	.0201	.0404	.103	.211	.446	.713	1.386	2.408	3.219	4.605	5.991	7.824	9.210
3	.115	.185	.352	.584	1.005	1.424	2.366	3.665	4.642	6.251	7.815	9.837	11.341
4	.297	.429	.711	1.064	1.649	2.195	3.357	4.878	5.989	7.779	9.488	11.668	13.277
5	.554	.752	1.145	1.610	2.343	3.000	4.351	6.064	7.289	9.236	11.070	13.388	15.086
6	.872	1.134	1.635	2.204	3.070	3.828	5.348	7.231	8.558	10.645	12.592	15.033	16.812
7	1.239	1.564	2.167	2.833	3.822	4.671	6.346	8.383	9.803	12.017	14.067	16.622	18.475
8	1.646	2.032	2.733	3.490	4.594	5.527	7.344	9.524	11.030	13.362	15.507	18.168	20.090
9	2.088	2.532	3.325	4.168	5.380	6.393	8.343	10.656	12.242	14.684	16.919	19.679	21.666
10	2.558	3.059	3.940	4.865	6.179	7.267	9.342	11.781	13.442	15.987	18.307	21.161	23.209
11	3.053	3.609	4.575	5.578	6.989	8.148	10.341	12.899	14.631	17.275	19.675	22.618	24.725
12	3.571	4.178	5.226	6.304	7.807	9.034	11.340	14.011	15.812	18.549	21.026	24.054	26.217
13	4.107	4.765	5.892	7.042	8.634	9.926	12.340	15.119	16.985	19.812	22.362	25.472	27.688
14	4.660	5.368	6.571	7.790	9.467	10.821	13.339	16.222	18.151	21.064	23.685	26.873	29.141
15	5.229	5.985	7.261	8.547	10.307	11.721	14.339	17.322	19.311	22.307	24.996	28.259	30.578
16	5.812	6.614	7.962	9.312	11.152	12.624	15.338	18.418	20.465	23.542	26.296	29.633	32.000
17	6.408	7.255	8.672	10.085	12.002	13.531	16.338	19.511	21.615	24.769	27.587	30.995	33.409
18	7.015	7.906	9.390	10.865	12.857	14.440	17.338	20.601	22.760	25.989	28.869	32.346	34.805
19	7.633	8.567	10.117	11.651	13.716	15.352	18.338	21.689	23.900	27.204	30.144	33.687	36.191
20	8.260	9.237	10.851	12.443	14.578	16.266	19.337	22.775	25.038	28.412	31.410	35.020	37.566
21	8.897	9.915	11.591	13.240	15.445	17.182	20.337	23.858	26.171	29.615	32.671	36.343	38.932
22	9.542	10.600	12.338	14.041	16.314	18.101	21.337	24.939	27.301	30.813	33.924	37.659	40.289
23	10.196	11.293	13.091	14.848	17.187	19.021	22.337	26.018	28.429	32.007	35.172	38.968	41.638
24	10.856	11.992	13.848	15.659	18.062	19.943	23.337	27.096	29.553	33.196	36.415	40.270	42.980
25	11.524	12.697	14.611	16.473	18.940	20.867	24.337	28.172	30.675	34.382	37.652	41.566	44.314
26	12.198	13.409	15.379	17.292	19.820	21.792	25.336	29.246	31.795	35.563	38.885	42.856	45.642
27	12.879	14.125	16.151	18.114	20.703	22.719	26.336	30.319	32.912	36.741	40.113	44.140	46.963
28	13.565	14.847	16.928	18.939	21.588	23.647	27.336	31.391	34.027	37.916	41.337	45.419	48.278
29	14.256	15.574	17.708	19.768	22.475	24.577	28.336	32.461	35.139	39.087	42.557	46.693	49.588
30	14.953	16.306	18.493	20.599	23.364	25.508	29.336	33.530	36.250	40.256	43.773	47.962	50.892

* Table VI is reprinted from Table III of Fisher: *Statistical Methods for Research Workers*, Oliver & Boyd Ltd., Edinburgh, by permission of the author and publishers.

For larger values of df, the expression $\sqrt{2\chi^2} - \sqrt{2(df) - 1}$ may be used as a normal deviate with unit standard error.

Index

Index

Standard deviation, 59
  of a binomial population, 64
  of a difference, 108
  of a set of ranks, 66
  of a sum of binomial variables, 135
Standard error, of a comparison, 163
  of a difference, 108
  of a difference between two means
    when observations are independ-
    ent, 147
  of a difference between two means
    when observations are paired,
    152–154
  of estimate, 79–81, 83
  of a mean, 105–107
  of a proportion, 122
Standard score, 72–73
Statistic, 8, 61
Statistical inference, 8
Stimulus variables, 12
Systematic errors, 207–208

T

t, distribution of, 142
  test of significance of a comparison,
    165
  test of significance for a correlation
    coefficient, 189
  test of significance of the difference
    between two means, 147–148,
    152–154
  test of significance of a mean, 142–
    143
  test of significance of a rank-order
    correlation coefficient, 181
Test of significance, in the analysis of
    variance, 160–161
  chi square, 173, 175–176
  of a comparison, 164–165
  of a contingency coefficient, 188–
    189
  of a correlation coefficient, 189
  of a correlation ratio, 186–187
  of the difference between two
    means, 147, 152–154

Test of significance (*continued*)
  normal-distribution free, 208–209
  one-sided, 149
  of a phi coefficient, 182
  of a point-biserial coefficient, 184
  power of a, 150–151
  randomization, 213–215
  of rank-order correlation coefficient,
    181
  of a treatment mean square, 160–
    161
  two-sided, 149
Test of technique, 212–213
Two-sided test, 149
Type I error, 149
Type II error, 149

U

Uniform distribution, 33
Univariate distribution, 25–27

V

Validity coefficient, 203–204
Variables, behavioral, 11–12
  binomial, 117
  continuous, 16–17
  definition of, 11
  discrete, 16–17
  linearly related, 76–78
  organismic, 12
  psychological, 11–12
  stimulus, 12
Variance, 58, 141
  of a binomial population, 64
  of a difference, 107–108
  of a difference between two means,
    145
  distribution of, 109–110
  of a mean, 105–107, 141
  of a population, 102
  and random errors, 196
  of a set of ranks, 65
  of a sum, 109
  of a sum of binomial variables, 121